Technology Commercialization

RUSSIAN CHALLENGES, AMERICAN LESSONS

Committee on Utilization of Technologies Developed at Russian Research and Educational Institutions

Office of International Affairs
National Research Council

Russian Academy of Sciences

NATIONAL ACADEMY PRESS
Washington, D.C. 1998

NATIONAL ACADEMY PRESS · 2101 Constitution Ave., N.W. · Washington, DC 20418

NOTICE: The project that is the subject of this report was approved by the Governing Board of the National Research Council, whose members are drawn from the councils of the National Academy of Sciences, the National Academy of Engineering, and the Institute of Medicine. The members of the committee responsible for the report were chosen for their special competencies and with regard for appropriate balance.

This report has been reviewed in draft form by individuals chosen for their diverse perspectives and technical expertise, in accordance with procedures approved by the NRC's Report Review Committee. The purpose of this independent review is to provide candid and critical comments that will assist the institution in making the published report as sound as possible and to ensure that the report meets institutional standards for objectivity, evidence, and responsiveness to the study charge. The review comments and draft manuscript remain confidential to protect the integrity of the deliberative process. We wish to thank the following individuals for their participation in the review of this report:

James F. Lardner (Deere & Company, ret.)
Lawrence Hodges (Technical Affairs Consultant, Ltd.)
Earl H. Dowell (Duke University)
John E. Halver (University of Washington)
William Gordon (Rice University)

While these individuals have provided constructive comments and suggestions, it must be emphasized that responsibility for the final content of this report rests entirely with the authoring committee and the institution.

Library of Congress Catalog Card Number 98-89705

International Standard Book Number 0-309-06194-6

A limited number of copies of this report are available from:

Division on Development, Security, and Cooperation
National Research Council
2101 Constitution Avenue, NW FO2060
Washington, DC 20418
Tel: (202) 334-2644

Copies of this report are available for sale from:

National Academy Press
2101 Constitution Avenue, NW Box 285
Washington, DC 20055
Tel: 1-800-624-6242 or (202) 334-3313 (in the Washington Metropolitan Area).

Printed in the United States of America.

LIST OF COMMITTEE MEMBERS AND STAFF

National Research Council Committee

ALVIN TRIVELPIECE (*Chair*), Oak Ridge National Laboratory
ALEXIS CLARE, Alfred University
W. MARK CROWELL, North Carolina State University
RICHARD DULIK, Covington & Burling
ALEXANDER MACLACHLAN, E.I. duPont de Nemours & Company (retired)
DAVID MCNELIS, Research Triangle Institute

Russian Academy of Sciences Committee

NIKOLAI LAVEROV (*Chair*), Russian Academy of Sciences
SERGEI IVANTCHEV, Plastopolimer Research and Production Association
YURI LEBEDEV, Ministry of Science and Technology
VLADIMIR LITVINENKO, St. Petersburg State Mining Institute
VLADIMIR MESHCHERYAKOV, Russian Agency for Patents and Trademarks
ALEXANDER OZERIN, Institute of Synthetic Polymer Materials
ALEXANDER SIMINOV, Karpov Physical Chemistry Institute

Staff

GLENN SCHWEITZER, Director, Office for Central Europe and Eurasia, NRC
STEPHEN DEETS, Office for Central Europe and Eurasia, NRC
KELLY ROBBINS, Office for Central Europe and Eurasia, NRC
YURI SHIYAN, Dept. of International Affairs, Russian Academy of Sciences

The National Academy of Sciences is a private, nonprofit, self-perpetuating society of distinguished scholars engaged in scientific and engineering research, dedicated to the furtherance of science and technology and to their use for the general welfare. Upon the authority of the charter granted to it by Congress in 1863, the Academy has a mandate that requires it to advise the federal government on scientific and technical matters. Dr. Bruce M. Alberts is president of the National Academy of Sciences.

The National Academy of Engineering was established in 1964, under the charter of the National Academy of Sciences, as a parallel organization of outstanding engineers. It is autonomous in its administration and in the selection of its members, sharing with the National Academy of Sciences the responsibility for advising the federal government. The National Academy of Engineering also sponsors engineering programs aimed at meeting national needs, encourages education and research, and recognizes the superior achievements of engineers. Dr. William A. Wulf is president of the National Academy of Engineering.

The Institute of Medicine was established in 1970 by the National Academy of Sciences to secure the services of eminent members of appropriate professions in the examination of policy matters pertaining to the health of the public. The Institute acts under the responsibility given to the National Academy of Sciences by its congressional charter to be an adviser to the federal government and, upon its own initiative, to identify issues of medical care, research, and education. Dr. Kenneth I. Shine is president of the Institute of Medicine.

The National Research Council was organized by the National Academy of Sciences in 1916 to associate the broad community of science and technology with the Academy's purposes of furthering knowledge and advising the federal government. Functioning in accordance with general policies determined by the Academy, the Council has become the principal operating agency of both the National Academy of Sciences and the National Academy of Engineering in providing services to the government, the public, and the scientific and engineering communities. The Council is administered jointly by both Academies and the Institute of Medicine. Dr. Bruce M. Alberts and Dr. William A. Wulf are chairman and vice chairman, respectively, of the National Research Council.

CONTENTS

PREFACE

In February 1997, Russian Deputy Prime Minister Vladimir Fortov, who subsequently became the Minister for Science and Technology and then Vice President of the Russian Academy of Sciences, suggested that the National Academy of Sciences and the Russian Academy of Sciences cooperate in helping to prevent the deterioration and improve the utilization of the industrial research base of Russia. Following discussions between representatives of the two academies, the National Academy of Sciences, joined by the National Academy of Engineering and acting through the National Research Council, agreed with the Russian Academy of Sciences to organize consultations between American and Russian specialists with practical experience in facilitating the commercialization of technologies in the two countries. These consultations, supported by internal funds available to the National Research Council and the Russian Academy of Sciences, took place primarily during a visit to Russia by the American delegation in November 1997 and a visit to the United States by the Russian delegation in March 1998. (The itineraries for these visits are in Appendix N.) Discussions during the visits touched on a small, but highly relevant, fraction of technology commercialization experiences in each country. This report reflects the issues that arose during these consultations and visits.

The specialists from both countries recognized the vast differences in the ways technologies are commercialized in the United States, with its thriving economy and robust industrial base, and in Russia, where economic conditions are weak and unstable and industrial production has declined to less than 25 percent of its level seven years ago. Nevertheless, in its move toward a market economy, Russia could benefit from American experiences as it considers a policy framework and programs to facilitate development of technologies with economic potential and to introduce them in the Russian and international markets.

The specialists did not attempt to predict the future of industrial development in Russia; rather, they assumed that industrial production would remain weak for the next several years and that most sectors of Russian industry would not be in a position to provide major financial support for R&D activities. Thus, their challenge was to identify approaches in the United States, Russia, and other countries that could preserve and even strengthen applied research capabilities under difficult economic constraints.

In some areas, such as oil, gas, and other natural resource sectors, Russian industry is in a position to provide financial support for R&D activities. Western firms are making significant investments in other areas, such as aerospace and a few other advanced technology sectors. And in still other areas, small, private innovative firms have appeared, often established as spin-off enterprises from large public-sector R&D facilities. Activities in all these areas represent profitable endeavors for some Russian R&D organizations. In addition, a few Russian groups have continued to provide technologies for the Russian military sector, albeit at a greatly reduced level. However, most applied research facilities have fallen on difficult times: laboratories are empty, the scientific work force is aging and underemployed, and interested customers are few in number.

Against this background, the American and Russian specialists present their individual views concerning problems, opportunities, and relevant experiences. They recognize that no generic solution to the various problems confronting Russian research organizations exists. Therefore, they offer a variety of ideas to help Russian R&D groups survive—and even thrive—in the current environment.

ACKNOWLEDGEMENTS

The American and Russian committee members wish to thank all those individuals who contributed to the success of this project. The many visits to institutes and government agencies in both countries were invaluable, and the committee members are grateful for the openness and hospitality which were displayed at each meeting. Yuri Shiyan deserves special recognition for his tireless efforts, and Carol Flath superbly served as translator during the Russian committee's visit to North Carolina.

This report has been reviewed in draft form by individuals chosen for their diverse perspectives and technical expertise, in accordance with procedures approved by the NRC's Report Review Committee. The purpose of this independent review is to provide candid and critical comments that will assist the institution in making the published report as sound as possible and to ensure that the report meets institutional standards for objectivity, evidence, and responsiveness to the study charge. The review comments and draft manuscript remain confidential to protect the integrity of the deliberative process. We wish to thank the following individuals for their participation in the review of this report: James F. Lardner (Deere & Company, ret.), Lawrence Hodges (Technical Affairs Consultant, Ltd.), Earl H. Dowell (Duke University), John E. Halver (University of Washington), and William E. Gordon (Rice University). While these individuals have provided constructive comments and suggestions, it must be emphasized that responsibility for the final content of this report rests entirely with the authoring committee and the institution.

PERSPECTIVE FROM A UNIVERSITY WITH AN INDUSTRY-FUNDED RESEARCH PROGRAM

Alexis G. Clare
Alfred University

INTRODUCTION

The New York State College of Ceramics at Alfred University is both a teaching and a research institution. Unlike research at many other institutions, the college's research is, in large part, industry funded rather than federally funded. This paper describes the nature of much of that industry funding, how it is obtained, and how consortia can be formed to leverage funds for high-risk or pre-competitive research. In recognition of the severe and daunting financial cutbacks that our Russian colleagues are experiencing, this paper describes some of the recent steps that my institution, as a state college, has taken to counterbalance potentially serious cuts in state funding of its academic program.

WHAT IS THE NEW YORK STATE COLLEGE OF CERAMICS?

The New York State College of Ceramics at Alfred University is a statutory college of the State University of New York (SUNY), which originally was established by the state legislature in 1900. SUNY long has been hailed as one of the more heavily subsidized state university systems and as one of the largest such systems in the country. Because SUNY, like other state universities in the United States, usually subsidizes tuition for students, it is advertised as offering an excellent education at a reasonable price. Its 64 separate campuses located throughout the state report to a central administrative body called SUNY Central. Some of these campuses are largely devoted to vocational education and offer two-year associates degrees; others grant undergraduate and graduate degrees. In all SUNY serves approximately 40,000 students.

Five statutory colleges belong to the SUNY system. Housed at private educational institutions, statutory colleges were established by the state legislature to provide specialized education related to the industries in the geographical area where they were located. For example, the College of Veterinary Medicine at Cornell University is located in a heavily agricultural area. The New York State College of Ceramics (NYSCC) at Alfred University was established to educate engineers and artists for the local ceramics industry. The private school in which NYSCC is housed is responsible for the day-to-day administration of the college program and for providing student services as well as accessory tuition for courses in subjects not taught in the college (liberal arts, business, natural science, and so on). Students at NYSCC pay substantially less tuition than students in other disciplines at Alfred University.

The reputation of NYSCC in both ceramic engineering and art is without equal. The college graduates one-third of the ceramic engineers in the United States. It offers three B.S. degrees, three Masters Degrees, and two Ph.D. degrees; the Ph.D. in glass science is offered at only two other institutions—one in Sheffield in the United Kingdom and the other in St. Petersburg, Russia. All of the aforementioned degrees are considered by academics and industrialists to be the best offered in their fields.

INDIVIDUAL INDUSTRIALLY FUNDED RESEARCH PROJECTS

Alfred University faculty members traditionally have conducted a larger proportion of research under industry contracts than under government grants. Compared to government grants, industrial contracts are of shorter duration with lists of well-defined deliverables and milestones toward meeting project goals. With industry contracts, considerable flexibility in intellectual property policies usually is required. Publication of results can be subject to veto or delay, depending on patent status and type of research. And because industry contracts entail confidentiality agreements and are often short in duration, they are not conducive to graduate student-based research.

For these reasons, the performance of basic research under industry funding is difficult. However, Alfred University has circumvented problems with such research by making agreements with the sponsoring companies concerning publication of research results. According to these agreements, research pertaining directly to a sponsor's product or process may not be published, but results obtained during the course of the contract that do not compromise the sponsor's competitive market position may be published (subject, of course, to examination of the proposed publication by the sponsor).

Some faculty members believe that industrial research is the scientific equivalent of prostitution, while others thrive in the problem-solving environment. Some industry representatives have complained that university researchers are slow to respond, lack focus, and generally do not understand industry needs. These criticisms usually can be reduced when scientists spend leaves of absence in industry. To build industrial partners' confidence in their abilities, scientists initially might have to take less interesting or financially rewarding projects to get the industry "hooked" on the institution and on reliance on the scientists' advice and abilities. Lastly, to obtain industrial contacts it pays to advertise both by word-of-mouth and on the World Wide Web.

STATE-FUNDED RESEARCH FOR INDUSTRIAL DEVELOPMENT

The ability of the College of Ceramics to carry out industrial research has been recognized by the State of New York's Science and Technology Foundation. For ten years the Foundation has designated the College's Center for Advanced Ceramic Technology (CACT)—which focuses on processing, characterization, and modeling of high performance ceramics, telecommunications, photonics, bioceramics, and other ceramic-related technology—as a Center for Advanced Technology (CAT). A CAT focuses on a high-technology field pertinent to faculty talents and supports economic development in New York State by (1) developing potentially commercializable ideas, (2) providing small companies with analytical facilities and problem solving assistance, and (3) helping larger companies with high-risk but potentially lucrative technologies. When a CAT receives $1 million in external funding, of which 60 percent must come from New York State companies, the New York State government will provide $1 million in matching funds. These funds are carefully reviewed every year.

In a growing market economy, virtually the only way to persuade government to invest in research is to show that the research will create jobs and increase the profitability of local firms. Testimonials from companies assisted by research institutions as well as a clear plan for leveraging government investments are extremely helpful in this regard. Documentation of the number of jobs created or saved, the amount of profit increase, and the level of production efficiencies achieved is all-important.

NSF INDUSTRY/UNIVERSITY CENTERS FOR RESEARCH

More than ten years ago, the National Science Foundation (NSF) established the Industry/University Cooperative Research Centers Program (IUCRC) to increase the participation of industry in research in universities and colleges. The idea was to establish industrial research centers at appropriate university sites to carry out pre-competitive research funded principally by industry and facilitated by a small contribution from NSF.

The philosophy behind these centers is that to remain current in the marketplace, companies must conduct a certain amount of pre-competitive research to ensure (1) a firm basis for development of novel ideas in a pre-competitive stage that might lead to profitable new products, (2) a logical approach to common manufacturing problems, and (3) the availability of reliable data for input into modeling sequences used in manufacturing. Such research typically is deemed important to the knowledge base of the company but is either too expensive for the company to conduct on its own, too high-risk, or too low a priority to command resources, which instead are diverted to immediate production problems and product development. It is our experience that fundamental studies can be financed by industry, provided that the studies are relevant to a current commercial product or process, are possible precursors to new commercial products or processes, or are funded in a leveraged fashion.

In the United States, approximately 50 centers have been established through the NSF program. Each center has a membership fee (usually about $30,000 to $50,000 per year), part of which may take the form of "in-kind" services, such as the provision of laboratory space, consumables, equipment, or personnel time. Member companies meet twice a year and receive reports on on-going research and new project proposals. They vote to select projects and fund them from the resources provided by the membership fees.

One of the first centers in the NSF program was established at Alfred University for research in the area of glass. The Center for Glass Research (CGR) now boasts 26 members from all of the major glass industries, along with their suppliers, and an annual budget approaching $1 million. The success of the CGR can be attributed to several factors. First, the NYSCC has a very long tradition of industry-sponsored research and has a reasonable understanding of industry's needs. Second, the CGR members have a great amount of input into the projects and how they are conducted. Finally, the center gives all members, many of whom are competitors, opportunities to meet and discuss common problems and possible solutions.

THE INDUSTRY/UNIVERSITY CENTER FOR BIOSURFACES

One discovery of the IUCRC program is that what works well for a traditional manufacturing industry, such as the glass industry, does not necessarily work well for other, newer industries. The biotechnology industry, for example, is much more diverse than the glass industry and much more complex in terms of intellectual property protection, government regulations, and development time lines. The NSF Industry/University Center for Biosurfaces (IUCB), therefore, has been less stable than the CGR.

The IUCB was established at the University of Buffalo. Its membership has fluctuated throughout its existence. Approximately six years ago, NYSCC joined the IUCB as a satellite center to provide a perspective in ceramic materials; the University of Memphis joined the IUCB to provide expertise in flow mechanics in cardiovascular systems. Last year, a third satellite with expertise in cardiovascular systems, at the University of Miami, also became part of the IUCB.

With 11 members, the IUCB is much more diverse in interests than the CGR, although its annual budget of $250,000 is considerably smaller. Its struggle to maintain membership is explained by the fact that many biotechnology firms are small companies for whom the membership fee usually represents a significant portion of their overall budget. These companies also have highly directed research needs. Some members of the IUCB are larger companies that currently have no products in the biotechnology area but that might wish to examine potential new product lines or the interaction of biological systems with their products for health and safety reasons. When there is little overlap between members' needs, consortium research is difficult. In addition, many companies are not satisfied with the IUCRC program's provisions on patent coverage and royalty-free licenses. These companies are inclined to pursue developments in-house, where they can keep discoveries more secure. Finally, small high-tech companies are far more competitive than larger companies and less inclined to pool resources and share ideas. Despite these problems, the IUCB continues to grow, albeit at a slower rate than the CGR.

Several conditions are needed to make a consortium like the IUCRC work well. First, the members need a common material or process about which they have shared concerns. All companies must pay the same membership fee and enjoy the same rights and privileges, regardless of size. Such equality can present a problem if companies considerably different in size wish to join the consortium. However, a policy of equality must be maintained to avoid the creation second-class members. Third, the institution's scientists must be prepared to listen to and work with the industry members; otherwise, their interaction is not as useful or productive as it might be. Finally, both industry and institution participants must understand the benefits of consortium research and should use the consortium as a springboard for further interaction.

While consortium-based industrial research is economical, in a nascent market economy the difficulties of creating a consortium can be severe. These difficulties might be avoided by consortia formed to focus on a very specific problem or to finance a specific invention with applications in two or more non-competing industries. The Whiteware Research Center at Alfred University represents the former type of consortium. It was formed in response to a production problem experienced by a local dinnerware company. The company, delighted with the response to its needs, contacted a number of other, similar companies to form a consortium. Alternately, a scientist might be able to initiate a mini-consortium to finance work on an invention that is ready for the pre-development stage and that may be of use to industry. This type of consortium works particularly well if the research serves two or more entirely different purposes and if non-competitive industries can use the invention specifically for their application. For example, a researcher could invent a non-stick coating which might be applied to windshields, sanitary-ware, industrial machinery, and cookware. Each industry making these products might adapt the invention to its specific need and apply for an application-specific patent.

STATE-FUNDED EDUCATION

When a new governor of New York was elected about two years ago, the generous budget for the state university system of which NYSCC is a part was threatened, and all of the SUNY units faced severe cutbacks, downsizing, and tuition increases. The statutory colleges were in a particularly precarious position because their cost to educate a student appeared considerably higher than that of a regular SUNY campus. Although NYSCC's research program enjoyed a large proportion of funding by industry, it too was put at risk because of its interdependence with the academic program.

To survive, NYSCC needed to respond quickly to the new fiscal environment. The school cut its budget by downsizing through natural retrenchment—for example, leaving positions open and those vacated by retirements unfilled. It also mounted an intense student recruitment campaign to make up the shortfall in funds with tuition dollars, and it achieved greater efficiency through economizing in programmatic offerings. At the same time, the school expanded its degree offerings to remain attractive to students and to offer commercially relevant programs.

While the financial problems are not completely resolved, the College's decision to be proactive rather than reactive appears to have been wise. Crises have a way of forcing changes in strategies and in expenditures—changes that are not altogether bad—although at the time they can be frightening. Sometimes the most advantageous moves—in this case, the expansion of course offerings—can be counter-intuitive.

SUMMARY

Success in industry-sponsored research requires flexibility in intellectual property policy (in accordance with any restrictions imposed when federal funding is involved) and understanding of industry needs. Such research must be fast, focused, and relevant; and scientists must listen to their industry counterparts and not assume that they, as scientists, know what is best for industry.

Research consortia need to have common ground and are easiest to create with large, well-established industries that are willing to interact with one another. Small consortia appear to work best for more highly focused research. In today's economy, probably the only way to obtain industry funding of basic research is through consortia of large companies desirous of carrying out pre-competitive research.

A proactive response to potential cuts is more beneficial than a reactive response, although the best response may not be the intuitive one. An analysis of the responses of other institutions in similar situations can be helpful. There is an old saying—"if you want to make an omelet you have to break some eggs." This is tough if you're an egg; but being in an omelet is more constructive than being an unbroken egg in a refrigerator full of bad eggs.

COMMERCIALIZING UNIVERSITY TECHNOLOGY

W. Mark Crowell
North Carolina State University

It would be presumptuous at best to expect any American specialist to develop recommendations for Russian institutions and officials on the basis of a cursory and rapid review of technology development activities in Moscow and St. Petersburg. Rather, the challenge is to focus on U.S. experiences that may have relevance to the Russian situation and to highlight approaches that may be helpful to Russian institutions seeking to increase their technology commercialization activities. These experiences and approaches are the context for this paper.

Extensive transfer of technology from university and nonprofit laboratories to the commercial sector is a relatively recent phenomenon in the United States. Although a handful of universities have been involved in technology transfer for many years, widespread university activity in technology commercialization began only with the passage of the Bayh-Dole Act in 1980. This legislation enabled universities, nonprofit research institutions, and small businesses to own and patent inventions developed under federally funded research programs. One purpose of the act is to provide incentives for universities to patent and commercialize their research discoveries and for industry to make the long-term, high-risk investments at universities and other institutions that are necessary for product development and introduction.

The Bayh-Dole Act has served as a catalyst for the development of patenting and technology transfer as new university functions, for the emergence of a new discipline of professional university technology management, and for the encouragement of greatly expanded university interest in economic development. Before passage of the act, fewer than 250 patents were issued to U.S. universities each year; and university research discoveries were seldom commercialized for the public's benefit. Today, U.S. universities obtain an average of almost 1,500 patents per year. Moreover, more than 200 universities are engaged in technology transfer, eight times the number in 1980. These universities have one or more specialists who are members of the Association of University Technology Managers.

This paper reviews salient points of the Bayh-Dole Act and focuses on those provisions that have been particularly helpful in promoting university involvement in technology transfer. A discussion of the current practice of university technology transfer emphasizes provisions in university intellectual property policies that, with the act, serve to promote and provide incentives for faculty and institutional patenting and licensing activity. Training activities and programs available to university technology transfer professionals are summarized. Finally, statistics reflecting the growth and sophistication of university involvement in technology commercialization are provided.

THE BAYH-DOLE ACT

The Bayh-Dole Act and its amendments provide the foundation on which current university technology transfer activity is based. As noted, the act was passed in 1980 (Public Law 96-517). It was amended in 1984 (Public Law 98-620). The following aspects of the legislation are notable:

- Bayh-Dole provisions apply to all inventions conceived or first introduced into practice as a result of a project funded either in whole or in part by the federal government.
- Universities must report each new invention to the sponsoring government agency within two months of disclosure of the invention to the university.
- The university must decide whether it wishes to retain title to the invention within two years of reporting the invention to the sponsoring agency. This time is shortened if a publication has triggered the one-year grace period for patent protection, as provided in the U.S. law. If this period has been triggered, the university must decide whether to retain title to the invention at least sixty days before the end of the grace period.
- Within one year of electing to retain title, the university must file a patent application for the invention. Within ten months of filing a U.S. patent application, the university must report to the government sponsor whether it wishes to file foreign applications. If the university fails to proceed on these fronts, the government may proceed with such filings on its own behalf.
- The federal government is given a nonexclusive, royalty-free, irrevocable license to use the invention to which the university retains title or have the invention used on its behalf.
- Any licensee holding an exclusive license for sales of products in the United States must substantially manufacture the product in the United States. This rule can be waived by the sponsoring agency if the university can show that a reasonable effort was made to find a company that would manufacture the product in the United States.
- Universities must give preference to small businesses in their licensing activities. Diligence is required, however: the small firm must be capable of properly developing the invention. A large company that has provided some

of the research funding leading to the invention may become an exclusive licensee.

- Universities cannot assign their rights in inventions to third parties, with the exception of firms hired by the universities to manage patent activities on their behalf.
- The inventor of a licensed invention must receive a share of any royalties received by the university as a result of the license agreement. After providing the inventors' share and covering expenses, the remaining income must be used by the institution to support research or education.
- March-in rights are retained by the federal government—that is, the government can require the university to grant a license to a third party if the invention has not been properly developed or commercialized within a reasonable time period.

Bayh-Dole has been extremely successful in promoting university and faculty involvement in patenting and licensing activities. The reporting and march-in requirements serve to ensure that the university, as owner of the intellectual property, exercises due diligence to identify, evaluate, protect, and commercialize intellectual property, thus helping to address issues related to public benefit from government research funds. By requiring that royalties be shared with inventors and by allowing universities to own and license inventions, Bayh-Dole encourages participation by faculty members and their institutions in patenting and licensing activities. By allowing for exclusive licenses, the act rewards the high-risk investment that industry must make to develop technology and pursue product development. Finally, by requiring a preference for small business, the act recognizes the importance the government has placed on the economic development potential of university inventors' involvement in technology transfer.

TECHNOLOGY TRANSFER WITHIN U.S. UNIVERSITIES

With the passage of Bayh-Dole and the introduction of patenting and licensing functions within research universities, a new discipline—university technology management—emerged almost overnight. Most universities involved in technology transfer created institutional policies for intellectual property, established patent committees, and adopted other mechanisms to assist the development of management structures and processes for carrying out the new activities. The Bayh-Dole Act, with its clear and rational provisions, provided a good framework for the development of university policies that deal with ownership of inventions, royalty-sharing and distribution, freedom of publication, invention evaluation and management, and conflict of interest.

Ownership

Under Bayh-Dole, universities may not transfer to other institutions ownership of inventions developed in whole or in part with government funds. Accordingly, most universities have taken the position that they own all intellectual property, regardless of funding source, developed by their employees and students using their laboratory facilities and equipment or developed under a grant or contract they administered. The underlying rationale of Bayh-Dole with respect to this issue is that invention ownership is necessary to uphold other key principles, including preference for small business and incentives for participation in the patenting and licensing process. Other public policy issues may be important to the university in pursuing the commercial development of inventions not developed with federal funds. As with federally funded inventions, ownership is a key requirement to ensure that the university has the leverage needed to protect and pursue the technology transfer objectives it deems important.

Royalty-Sharing and Distribution

Bayh-Dole requires that royalties from licensing be shared with the inventors and that remaining income, after payment of expenses, be directed toward the support of research and education within the university. Recognizing the need to provide an incentive and reward structure for faculty to participate in the technology transfer process, many universities have included generous sharing formulae within their institutional patent policies. Typically institutions share 35 to 40 percent of royalties with inventors, and some share as much as 50 percent. As a further incentive, many universities allocate additional royalty shares to departments, inventors' laboratories, schools, or other entities within the universities. This approach provides incentives for these entities to support and promote technology transfer, and it ensures compliance with Bayh-Dole provisions requiring that royalties, after payments to inventors, be used to support research and education.

Freedom to Publish

Even with the advent of Bayh-Dole and intensified university interest in technology transfer, the basic reward structure within academic institutions remains traditional publications. The publication issue is important and challenging for many reasons. Faculty involved in patenting and technology transfer typically interact closely with industry, which often prefers to keep information as trade secrets for as long as possible. Further, both universities and their corporate partners fully understand the impact of premature publication on the availability of patent protection. Therefore, careful and realistic publication provisions are required in patent policies to protect the ability of the university to publish as it sees fit while providing reasonable and adequate safeguards for pre-publication review of proposed publications for patentable material.

Invention Evaluation and Management

With the increase in activity in patents and technology transfer, many universities have found it necessary to incorporate into their patent policies specific provisions dealing with the mechanism for properly assessing and managing university-owned inventions. Often this mechanism takes the form of a faculty-based intellectual property committee. Its purpose is to evaluate invention disclosures to the university and to assist the university technology manager in developing appropriate management plans for specific inventions. Some institutions have attempted to develop objective, quantifiable invention assessment instruments; the success of these instruments is the subject of serious debate and scrutiny. Regardless of the mechanism, assessment has become increasingly important because of its resource allocation implications. With increasing numbers of invention disclosures and patent activities and the high costs associated with such undertakings, initial assessments or "triages" of inventions are becoming important to maintain adequate resources for the pursuit of truly important or breakthrough discoveries.

Conflict of Interest

Increased involvement in technology commercialization has resulted in a heightened awareness of, and opportunity for, conflicts of interest. Many universities have addressed these conflicts within their intellectual property policies. Issues of importance include manipulation of research results for personal enrichment, inappropriate use of university-owned facilities and equipment, and improper influencing of graduate students to pursue research for profit rather than for knowledge. As universities have become involved in the creation of start-up companies to commercialize their technologies, often resulting in the acquisition of stockholdings by the university and the inventors, these issues have become more important and complex. Universities that have developed good conflict-of-interest policies often find aggressive technology transfer initiatives easier to pursue because they know the ground rules for use of research results, university-owned facilities and equipment, and graduate students.

TRAINING OF UNIVERSITY TECHNOLOGY MANAGERS

With the introduction of technology transfer as a new function within the university and the emergence of the new profession of university technology management, issues related to training and professional development have emerged. These issues fall largely within the domain of the Association of University Technology Managers (AUTM), an organization of specialists involved in university technology transfer from more than 200 educational institutions in the United States and Canada. In addition, some business schools are beginning to offer training in university technology transfer management, but these programs are rather scarce and typically are not involved in outreach to or continuing education for practicing university technology managers.

AUTM is a professional development organization known worldwide for its achievements in the teaching of technology transfer and intellectual property management principles and practices. Activities of AUTM include:

- Annual conferences, which typically attract almost 1,000 registrants and which offer 3 or more days of lectures, workshops, discussion and special interest groups, and networking opportunities.
- An annual basic licensing course and an annual advanced licensing course for newcomers and experienced professionals, respectively. The basic course is designed to teach the basic principles of negotiating technology license agreements to individuals with little or no experience in the field. The advanced course is aimed at bringing together experienced practitioners for exploration of specific topics related to the practice of technology transfer and professional development.
- Regional meetings are held each summer. These meetings are designed to bring smaller groups of technology transfer professionals together for more informal educational and networking opportunities. Preliminary discussions about the possibility of organizing a regional meeting for European or Asian members of AUTM are underway.
- Special training courses, such as the 1997 conference organized in Amsterdam with European organizations on "U.S. Methods in Technology Transfer/Best Practices," are sometimes organized by AUTM on request. The Organization for Economic Cooperation and Development (OECD) is considering sponsorship of a similar conference in Russia with AUTM participation. The author is chairman of AUTM's International Members Committee and is participating in the preliminary discussions with the OECD.
- The *AUTM Journal* is published semi-annually and includes invited articles on various topics by experienced practitioners.
- The AUTM Technology Transfer Practice Manual was published by AUTM as a "how to" reference guide for university technology transfer offices. The manual is a massive three-volume set that includes recommended basic and standard agreements for different types of

technology transfer deals (for example, licenses, options, research agreements, confidentiality agreements). Diskettes that contain sample agreements are provided with the manual.

TRENDS IN U.S. UNIVERSITY TECHNOLOGY TRANSFER

In the wake of the Bayh-Dole Act and the growth and maturation of university technology transfer practice in the United States, with the development and implementation of informed university policies addressing various matters pertaining to technology transfer, and with the emergence of AUTM, what has been achieved?

An analysis by AUTM (*AUTM Licensing Survey FY 1991 – FY 1995*) has shown that the licensing of university inventions adds more than $21 billion to the U.S. economy and supports an estimated 180,000 people each year. In 1995 alone, university technology licensing led to the formation of 223 new companies. AUTM reports that since 1980 university technology transfer has led to the formation of 1,633 new companies. Statistics from a AUTM survey indicate the extent of U.S. university involvement in technology licensing:

Activity	FY 1995	Cumulative % Change FY 1991-95
Invention Disclosures	7,427	+29%
Total U.S. Patent Applications Filed	5,100	+127%
New U.S. Patent Applications Filed	2,373	+53%
Licenses and Options Executed	2,142	+66%
Licenses and Options Generating Royalties	4,272	+72%
Gross Royalties	$274 million	+108%
Legal Fees Expended	$60 million	+82%
Total Sponsored Research Expenditures	$17,212 billion	+29%
Research Expenditures, Federal Funds	$11,381 billion	+23%

The dramatic increase in U.S. technology transfer and business development since the advent of Bayh-Dole, particularly during the recent five-year period reported by the AUTM survey, reflects a process that is working well and that is meeting many university technology transfer and economic development objectives set by government and university officials. The U.S. experience has focused on the development of a national framework and uniform policy within which technology transfer can be pursued; the emergence of institutional patent policies designed to manage in a rational fashion the many issues that arise within a university setting; and the emergence of a profession, with its own organization and journal, to influence expanding activities in this sphere. Certain aspects of these experiences perhaps could be adapted to the legal, policy, and institutional framework within which Russian institutions must operate, thereby helping to influence positive outcomes for those interested in commercializing technology developed at research and educational institutions.

LEGAL ISSUES OF SPECIAL CONCERN TO TECHNOLOGY COMMERCIALIZATION

Richard Dulik
Covington and Burling

INTRODUCTION

Committee Member: You have succeeded in setting up a viable technology company here in Russia. How did you do it?

Russian Entrepreneur: We collected some of the best people in our area of technology and started a company. Now we do contract work for many foreign companies. The hardest part was structuring the company to avoid the tax situation here. We could not just set up the company and pay our employees from the proceeds of the contracts. We would lose most of the money to taxes. Luckily, one of the companies we contract with has very clever lawyers.

Committee Member: What did you do?

Russian Entrepreneur: We had to set up a small foreign company. It is just a shell. We also made sure that all of our employees had credit cards. Our foreign company is paid under our contracts, and we pay our employees by having the foreign company deposit money in each employee's credit card account. Our company here in Russia gets just enough money to pay for the necessary expenses. This arrangement allows us to avoid most of the taxes. Otherwise, we might be losing 70 percent of our income to taxes.

We have no way of determining if this story is true. Nevertheless, it illustrates some of the legal challenges faced by the institutes and technology start-up companies in Russia today. Obviously, the story is about a clever way that one company found to reduce the amount of taxes it was paying. However, it is also about the need to find foreign partners who will tolerate such solutions and for legislation to support research institutes and technology start-up companies.

Companies, wherever they are located, will do what they can to reduce their tax burdens. This is simply good business practice. In the United States, there is a never-ending search by both companies and individuals for loopholes in the tax laws. However, a company that has to do business in ways as complex as those described by the Russian entrepreneur will have trouble finding foreign partners because the possibility that something will go wrong increases the risks of the foreign partners. All other things being equal, potential partners will prefer the least risky situation; and the least risky situation may not be working with a company operating in Russia. Moreover, potential partners may fear the high visibility of Russian research institutes and technology start-up companies as potential sources of foreign exchange. In Russia, where it is difficult to effectively tax significant portions of the economy, these institutes and companies may become potential targets for more than their share of taxes and regulation.

The story also illustrates some desirable developments in Russia. Many Russians have adapted quickly to a free market system and are using great creativity to work around legislation that may not be encouraging business development. In addition, much Russian science is of excellent quality and is highly competitive in the international arena. Unfortunately, to a significant extent, Russian science is not currently competitive in Russia itself because of the lack of infrastructure to convert ideas into saleable products.

This paper will further examine the above points while discussing the legal environment in which research institutes and technology start-up companies are attempting to commercialize Russian technology.

TAXATION

In addition to illustrating the general environment in which Russian companies operate, the story related in the Introduction specifically highlights tax questions. There appears to be great disagreement among Russians themselves as to the actual tax burden faced by research institutes and technology start-up companies. In the past, there was always great uncertainty about taxes, and the tax situation remains in flux as new legislation is introduced.

Regardless, some general comments can be made in light of American experience. Government tax policies have enormous impact on the viability of business investments. Tax environments that are harsh put exceptional pressure on R&D spending because much of the return on investment will come in future years. When tax rates are high, the time value of money makes short-term high returns much more valuable than long-term low to moderate returns. In Russia, we heard many times that rich "New Russians" will not invest in R&D because much higher short-term returns are available from other enterprises.

The federal government has tried various tax incentives to encourage companies to invest in R&D in the United States. Many state governments have special programs to subsidize start-up companies in various ways and provide tax advantages. These programs are often successful in encouraging companies to operate in those states. The philosophy behind these programs is that the government is willing to forego some short-term income from taxes and fees to create long-term growth that will produce jobs and strong, innovative companies that have a future.

Similarly, universities often enjoy tax advantages in the United States as nonprofit organizations, while government entities or quasi-government entities may not pay any tax at all or enjoy reduced taxes. It must be noted, however, that the U.S. tax code is notoriously complex and by no means a necessarily good model for Russia. In practice, determining whether a university must pay federal income taxes on income derived from certain activities may be quite difficult. For example, a university may have to pay income tax on activities that appear to be unrelated to the university's tax-exempt educational, charitable, or scientific purposes. An introduction to this complex subject can be found in Kertz and Hasson's article, "University Research and Development Activities: The Federal Income Tax Consequences of Research Subsidiaries and Joint Ventures," *Journal of College and University Law* (Vol. 13, No. 2).

The U.S. government does not tax itself; the income earned by government entities ultimately is controlled by the U.S. Congress and may or may not be spent by these entities. Naturally, this method of operation requires a complex budgeting process to ensure that entities are able to spend sufficient funds (whether derived from the U.S. Treasury or from earnings) to fulfill their assigned missions. It is unclear whether a similar and effective system is currently operating in Russia, given the many complaints that institutes had insufficient funds to regularly pay livable wages and that income from foreign sources, such as research grants, was subject to substantial taxes.

Start-ups and institutes may become targets in Russia for taxes because their incomes can be monitored by the government; the institutes, in particular, have less flexibility to creatively structure themselves to avoid taxes. The Russian government does not appear to have efficient mechanisms to tax personal income, and therefore, gains little income from the job creation process that is so widely encouraged in the United States.

The taxes and other financial burdens currently placed on institutes and start-ups in Russia is unclear. It therefore would be very useful to calculate the entire cost burden of government taxes and other fees under the existing and proposed tax legislation that would be faced by a hypothetical start-up company. Whatever the result, the tax and fee burdens must be manageable or institutes and start-ups will cease to exist. Then the government will lose both the short-term and potential long-term income.

One additional consideration may be unique to the current situation in Russia. Because of the elimination of a substantial portion of the infrastructure capable of developing and producing high-technology mass-market goods, the relatively sophisticated technology of research institutes must be marketed primarily overseas, a point discussed in greater detail below. Generally, a country's ability to produce technology products keeps pace with its ability to carry out research to develop those products. Since the Russian research capabilities currently cannot be used on a significant scale domestically, the institutes must market, to a large extent, outside Russia. As a result, they probably incur costs not borne by research organizations in other countries. Therefore, to be competitive, Russian institutes must bear lower taxes, fees and other burdens than those found in other countries, not higher as now appears to be the case.

CONTRACTUAL AND BUSINESS MATTERS

Many Russian institutes are conducting scientific research that is equal to or better than the work of their counterparts in other countries. Unfortunately, the production infrastructure—including established manufacturing companies, venture capitalists, and start-up companies that normally would be expected to convert scientific research into successful products—is only now beginning to develop in Russia. The political and economic changes during the last 10 years may have severely damaged Russia's scientific establishment, but they have virtually destroyed significant portions of Russia's industrial production capabilities, particularly in the area of high-technology goods. We frequently heard that many production facilities operating during the Soviet period had become antiquated and unable to compete directly against the more modern facilities found in many other countries. When thrust into competition with those more modern facilities, the Russian facilities had been forced to close since there was apparently insufficient investment capital available to fund modernization. Consequently in some industrial sectors both sophisticated new technologies and potential consumers of that technology are present but the production infrastructure is absent. Therefore, a current natural market for the scientific products of institutes and start-ups in these sectors is overseas—particularly in the advanced economies of the United States and Europe, where high-technology product development and production facilities are readily available.

The marketing of technology and products overseas is, however, a substantial task. For example, one Russian institute director described his institute's experiences with competitive contracts in the West. The institute had submitted a series of proposals, apparently in response to competitive Requests for Proposals (RFPs), but had won no new business. The officials of the institute were frustrated, felt that there was discrimination against Russian organizations, and resolved not to compete for foreign business again.

Russian institutes and start-ups must face many hurdles in entering the international marketplace. First, Russian institutes, perhaps due to a long history of responding to explicit government requirements for products or results, may not have much experience in producing proposals that compare favorably to those produced by companies in the West.

Second, according to many Russians, few Russian products are produced in compliance with recognized international product standards. We were told that many Russian products in certain fields were equivalent or superior to those marketed in Western Europe but that selling Russian products was virtually impossible because they did not comply with the very detailed European Union product standards. To compete internationally, Russian companies recognize that they must both understand the new standards and demonstrate that they are capable of producing products to those standards.

Third, Russian executives may have little experience in highly competitive environments, where companies expect to lose a substantial majority of contract competitions. The effort expended by companies on a particular proposal must take into account the chances that the company actually will win a contract. The company should not have high expectations of winning any particular competition.

Fourth, Russian companies in highly competitive markets may lack a thorough understanding of competitive pricing. Companies in these markets must understand both whom they are competing against and what the competitors are bidding. Many consulting firms can supply this type of information. For example, it is not uncommon in some industries for consultants to conduct "blind benchmarking studies" wherein competitors supply pricing information to consultants with the understanding and appropriate assurances that the sources of the information will not be identified. Russian companies must be prepared to obtain such information if they hope to succeed in competitive markets.

Fifth, few Russian companies leverage their purchasing requirements to obtain assistance and concessions from their suppliers. It would be appropriate for Russian companies that make substantial purchases of equipment or supplies from overseas suppliers to require suppliers to provide them with various forms of business assistance as part of a deal. For example, the Russian company, in return for placing a large order with an American computer manufacturer, could require the manufacturer to help the Russian company set up a modern procurement department.

Finally, the "command economy" mindset may have encouraged many Russian executives to adopt a passive operating style, whereby they expect opportunities to come to them. By contrast, American executives spend substantial amounts of time networking, marketing, visiting potential customers, and otherwise promoting their companies. Particularly in these early stages, as Russian companies emerge into the international marketplace, the leadership of the institutes and start-up companies must devote substantial time to visiting potential customers and seeking out new opportunities and applications. When funds are scarce, travel may appear to be an unnecessary expense, but it may more than pay for itself by generating opportunities for long-term business relationships.

Fortunately, the problems mentioned above are easily corrected. Over time, the executives of institutes and start-ups will gain experience with the competitive contracting system, international standards, and the art of marketing products and services. Also, relatively inexpensive training could be useful. Executives can be taught how to allocate resources efficiently, seek out market opportunities for their products or services, and write effective proposals.

INTELLECTUAL PROPERTY

Given the constant stream of controversies in the United States over the appropriate scope of intellectual property protection, the satisfaction of Russian specialists with current Russian intellectual property law was surprising. On the other hand, there was also uniform concern that Russia did not have effective mechanisms for enforcing intellectual property laws. In truth, such laws are inadequate by definition if they cannot be enforced. Attempts at enforcement usually bring to light defects in the underlying legal structure.

Russian specialists generally stated that the only way to enforce intellectual property laws was to bring suit. However, such action is rare because of its high costs and the uncertainty of the outcome. Therefore, intellectual property laws are widely ignored. Naturally, much depends on the nature of the technologies or products involved. Where the technologies are evolving rapidly or are easily concealed from reverse engineering, such as with some integrated circuits and complex industrial materials and processes, there is much less need to rely on the legal system for intellectual property protection. Products that are easy to duplicate, such as certain software, audio and video tapes, and other information products, are very vulnerable to piracy and generally must be protected through legal means.

With respect to enforcement of U.S. intellectual property laws, American specialists probably would make comments similar to those of Russians, although they would reach the opposite conclusion about the effectiveness of this enforcement. In the United States, enforcement is also primarily done through the courts, an expensive and uncertain process. For many smaller companies, intellectual property litigation is termed "Bet the Company" litigation because defeat will certainly drive the company into bankruptcy. Merely obtaining U.S. patent protection for a single invention can cost between $5,000 and $15,000, and litigation of patent infringement costs millions of dollars. In litigation, case-deciding decisions may be made by judges or juries with little understanding of the complex legal and technical issues involved, hence the great uncertainty regarding the outcome of many cases.

Why then, if the Russian and the American criticisms of approaches to enforcement are the same, do U.S. intellectual property laws serve as effective deterrents in the United States? It is probably because the prohibitive costs and uncertainty of American intellectual property litigation actually strengthen the impact of intellectual property laws.

The threat of a devastating lawsuit is always present. Despite the potentially serious consequences of losing, suing is culturally acceptable in the United States. Further, the rewards for winning or settling a lawsuit can be very large. U.S. intellectual property law may not only provide the winner with compensation for any damages suffered but also with substantial "bonus" money (punitive damages) that is intended to punish the loser for wrongdoing. This money provides not only the incentive to sue but the means to pay for the suit as many attorneys will accept a percentage of the winnings (contingency fees) as their sole compensation. Some other countries encourage suits by requiring the losing side to pay the legal costs of the winning side.

On the other hand, the threat of litigation encourages the vast majority of potential suits to be settled out of court through payments, cross-licensing, or other agreements. The uncertainty of litigation also encourages settlement because incorrect decisions can be made even in very clear cases, and appeals are as uncertain and expensive as the original suit. Thus, the mere potential for a suit is an incentive for companies to resolve their differences between themselves.

The possibility that a suit could bankrupt a company is never taken lightly. But the U.S. legal and business systems soften the impact of bankruptcy. In the absence of criminal wrongdoing, the individuals involved may simply find jobs elsewhere, while the company itself may be reorganized and remain in business.

Development of an effective legal system is important to the revival of the Russian economy but is likely to occur slowly. As a result, the market for the products of institutes and start-up companies lie outside Russia. Well-developed foreign legal systems can provide Russians with intellectual property protections when they market internationally. Contracts and companies may be structured so that they operate within Russia but rely only minimally on the Russian legal system.

CONCLUSIONS

Perhaps the single most important step the Russian government can take to support the research institutes is to clearly define the institute's position in modern Russian society. The leaderships of institutes cannot plan for the future when the ultimate existence of their organization remains in question. Current attempts to partially fund large numbers of institutes will only perpetuate the present situation, in which many highly educated and competent scientists are not paid enough by the government to survive and have no access to the equipment they need to perform useful research. Institutes that may have the capability to flourish as independent businesses need the opportunity to be able to do so, and institutes that have remained fully functional should, of course, be encouraged to continue with their work.

We learned that the Russian government is undertaking a program to certify institutes as legitimate research establishments. Such efforts undoubtedly will result in some dislocations of personnel, and programs must be developed to aid such persons. However, using some institutes merely as mechanisms for delivering partial financial support and social services to scientists is pointless and a serious waste of resources. Even direct financial aid to unemployed scientists, channeled perhaps through the Russian Academy, would be preferable because a clear distinction could be made between funds that are supporting actual research and funds that are providing social services.

The government must provide an environment in which well-managed institutes and start-up companies can flourish. The most difficult challenge may be to explicitly relax control of the institutes and the production facilities. The institutes, after all, were home to some of the Soviet Union's most closely guarded secrets. In modern Russia, the institutes are highly visible to the government and obvious targets for taxes and controls. Legislation that could affect new start-ups, particularly tax and import/export controls, should be specifically evaluated for its impact on investment, business development, and entrepreneurship in Russia.

As noted above, many of the institutes that we visited consider current Russian intellectual property laws adequate, but they add that the Russian legal system as a whole cannot be used to effectively protect intellectual property rights. The challenge is to develop a legal system that is accessible to individuals and small private entities and that can be seen to operate at least honestly if not predictably. This problem, which goes well beyond the scope of the committee's study of technology commercialization, should be considered well worth a separate study.

International contracting practices, business development, entrepreneurship, and standardization pose less serious problems. Facility in these areas can be taught. The solution, in one sense, is as simple as identifying appropriate people at the institutes to receive training and ensuring that they receive proper training. Because many scientists are not sure how they are expected to operate in the new environment and some, at least, hope for a return to the more structured environment of the past, the real challenge is to reorient and restructure the institutional leadership so that entrepreneurial personalities can rise to the top and lead their organizations into the future.

AN INDUSTRIAL PERSPECTIVE ON TECHNOLOGY COMMERCIALIZATION IN THE 1990S AND BEYOND

A. MacLachlan
E.I. duPont de Nemours & Company (Retired)

Many lessons can be gleaned from the struggles of industry, universities, and the federal government in the United States during the last decade to improve technology commercialization performance. The resulting adjustments of U.S. technology-oriented firms, although unlikely to be directly applicable to the situation faced by the Russian scientific enterprise as it tries to reorient itself toward the needs of a free enterprise system, may be adaptable to the Russian scene.

The struggles referred to above have unquestionably yielded fruit. Today U.S. companies generally are acknowledged to be quite competitive in world markets. They are considered innovative, and even in the eyes of their most stern critics largely they have overcome accusations that they were slow-moving, low-quality, high priced behemoths—products of a spoiled past when no real global competition existed.

This paper is written from the perspective of long-established companies, such as Dupont, GE, IBM, GM, and Merck, which to the outside world appeared to have the most acute problems meeting the demands of the new economic realities. However, the lessons these companies learned and the changes they adopted clearly are just as integral to the success of relatively new companies such as Intel, Compaq, and Microsoft.

What changes have improved technology commercialization? They seem to fall into two major categories: improved information processes—improved communication and sharing of information among all personnel involved in the commercialization process, and improved technology acquisition processes—improved understand of how to gain new technology from the "best" source(s) to reduce the risk, lower the costs, and decrease the time to convert new technology into products, processes, and services. To be sure, these categories overlap and may oversimplify the changes that U.S. companies adopted. Nevertheless, they provide insight into the key elements of the rejuvenated capabilities of modern companies.

IMPROVED INFORMATION PROCESSES

Improved communications and sharing of vital business information among all parties involved in the commercialization process appears to be the single most important change. Many books have been written on the subject.[1] According to most of these books, the "old" R&D department functioned independently within a company or business unit, largely deciding for itself which new products and processes the company or unit needed. Customers were involved only when prototype products were available, and manufacturing specialists then were expected to figure out how to make the product at a price and level of quality acceptable to customers. This portrayal appealed to readers not really familiar with the complex world of R&D within a company. For those in the real world, it often was viewed as wrong and too simplistic.

By the early to mid-1980s, and in many cases long before, the business and R&D leaders of most companies realized that something was wrong with their internal relationships. R&D was viewed as isolated, alarmingly independent, and very costly. It took some time, but events in world markets forced a rethinking of the role of R&D organizations within business structures. These events included a continuing loss of market share to foreign competition, recognition of poor product quality, excessively high prices, and slowness to respond to customer needs. Many internal and external surveys and studies clearly demonstrated erosion of profitability, poor use of capital, decreasing innovativeness, and many more indicators of noncompetitive business performance.

Once business and R&D managers finally acknowledged a mutual problem in understanding each other, they quickly recognized that R&D departments were not really in the mainstream of all informational input critical to a business. Many steps were taken to change this situation. All have the same overriding objective: to ensure that the R&D organization is fully integrated into the business "team" and feels intense responsibility for business success.

Today's successful businesses are run as teams. When planning technology strategies, these teams involve all the important functions (marketing, manufacturing, sales, finance, and R&D). Clearly, customers are not involved in all proprietary planning, but customer views are sought early in the technology planning stage to help business and R&D leaders decide whether to proceed in certain new directions. This input helps the business teams understand the customers' needs. The more trust that can be generated between customers and business planning teams, the better the overall result.

The involvement of R&D in business planning must be at all levels. It is not sufficient for just the management to be privy to business plans. Furthermore,

---

[1] See, for example, Philip A. Roussel et al., *Third Generation R&D,* Harvard Business School Press, 1991, and Michael L. Dertouzos et al., *Made In America,* M.I.T. Press, 1989.

when the pertinent research personnel are involved in business planning and exposed to all the information critical to the business, they are often in a much better position to realize when things are not going well with a technology development, or when new technology-based opportunities may be available. R&D personnel must visit customer sites, jointly plan and review status with marketing and manufacturing groups, and communicate effectively with and listen to the business leadership. No longer can the R&D function be referred to simply as those "scientists and engineers," but rather the personnel must be routinely thought of as valued participants in all aspects of business planning and investing.

Within Dupont, this new team approach has led to a virtual "explosion" of innovation around business needs. To name a few, these include new methods to manufacture polyester polymer that lowers energy costs, improves product yield and lowers investment; novel approaches to turn waste products with major environmental effects into valued new commodity chemicals for nylon processes; and innovative ways to heat polymer reactant streams with lower energy and more control to reduce investment and produce higher quality and in some cases new kinds of products.

Many "best business" practices have been devised to ensure the intimate involvement of R&D departments in all aspects of business management. One of the most valuable practices is structured progress reviews of each ongoing project for business team members. These reviews require collaborative decision-making about whether to proceed with new technology exploration or new and improved product and process developments. The goal of these reviews is to ensure that representatives of each function within the business understand the progress being made by R&D, give input and advice, and make adjustments in response to the new information, thereby minimizing the time and cost to complete projects as well as problems after the products are offered to customers. This "best practice" has many names, among them "Stage Gate" and "Product and Cycle Time Excellence." Other "best practices" include profit sharing to ensure that all functions within a business have a financial stake in success, the use of metrics to judge the impact of R&D performance on business success, and the provision of rewards and public recognition for exemplary performance.

Companies like Dupont, GE, and IBM historically had large basic research laboratories that operated independently of the individual business units, so the establishment of links to the business information stream has been particularly challenging. Some companies decided to give up their broadly based corporate research laboratories. Other companies, such as the ones mentioned above, retained these central organizations because of their powerful capabilities to hire the best technical people and achieve real technology breakthroughs in support of business objectives. Great success has been achieved. In Dupont, for example, the corporate research department is no longer viewed as an "ivory tower" isolated from business concerns, but instead as a key player in maintaining the company's competitive advantage.

In Dupont's case, one of the most important changes was a decision to involve all levels of the corporate research organization in business technology planning. Corporate personnel were expected to help identify areas of technology breakthrough in which one or more businesses could gain major competitive advantage. These were called commitments to "grand challenges" and have been extremely successful. One recent breakthrough was the development of an energy-efficient membrane process to recycle hydrochloric acid back to chlorine. Hydrochloric acid is a major by-product of almost every commodity chemical process, and its disposal is often a major cost and environmental problem. Other changes at Dupont included management exchanges with the business R&D organizations, the placement of corporate and business R&D personnel in the same buildings, and the establishment of a Corporate Technology Council, which is made up of leaders from all the company's R&D organizations and gives input and direction to the total corporate research effort. For the R&D organizations, the success of all these changes is evident in their increased budget and hiring allocations, their enhanced responsibility for certain technology areas, and their role in planning the use of the entire company's pilot plants.

IMPROVED TECHNOLOGY ACQUISITION PROCESSES

The second key change that has improved technology commercialization is acquisition of technology from the best sources available, sources that are both cost-effective and timely. A few years ago most U.S. and European companies were fairly self-reliant in developing technology and had been so for several decades. They created products and services from technology they had developed internally or from technology acquired technology in its early stages from inventors or other companies. Collaboration among companies or even with purveyors of technology was the exception rather than the rule.

In order to succeed in today's world of intense and highly capable competition, companies have become skilled at acquiring and commercializing technologies in new and quicker ways. Often these technologies are not exclusively theirs, but they are obtained from some outside source or co-developed with outside partners in order to save time and money. Today's competitors are very conscious of the time to go to the market place and the need to have better costs and performance than their competitors. Companies no longer believe with blind faith that research organizations are delicate flowers that must not be pressured nor required to justify their existence on a daily basis; nor do they believe all the new technology they need must be developed by their research organizations. In fact, many believe that the chief value of research organizations is their knowledge about where appropriate new technology might be purchased or about technology that might be developed in new partnerships.

For more than a decade, U.S. companies have worked hard to adjust to these new realities. Today, many would say that U.S. companies are ahead of much of their global competition. If so, it is not through brilliance, but because they have responded to a stark need with many years of often painful adjustments.

Research organizations are still viewed as vital but in a significantly different way than in the past. Companies expect them to be cost-effective in every sense of the word. In some companies, research organizations have been severely downsized and in some even eliminated. In the latter case, companies believe they can purchase the technology they need to support what they feel are their real strengths, which might be marketing, product design, or manufacturing. The companies that have downsized and reoriented their research organizations and expect these organizations to obtain technology at the lowest possible cost and at the fastest speed.

For this reason, companies no longer give unquestioned grant money to universities for good will purposes. Instead they expect *quid pro quo* arrangements. After their initial shock, universities long accustomed to unrestricted grants view this development as a positive one. It brings university professors and students closer to the real needs of industry and gives universities that had not benefited from grants an opportunity to compete for industry support.

Universities today rapidly are becoming integral to many companies' business strategies, and not just in the United States. All over the world, universities and companies are establishing research programs to augment corporate research organizations. Russian, Chinese, and Indian research institutes and universities, for example, have benefited from this new approach. The resulting benefits include access to outstanding research personnel and facilities, lower research costs, greater responsiveness to market needs, and greater knowledge of new markets. But in spite of this increasing trend towards partnerships, usually less than 1 percent of a company's R&D budget is spent at universities. Partnerships represent a potential growth area, provided the universities work diligently to understand business needs and modify their policies appropriately without compromising their fundamental missions.

Companies' new approach to technology acquisition is not confined to universities. Partnering is taking many forms. The most common form is partnering with other companies, both large and small. Often large powerful companies with global marketing organizations will collaborate with smaller more entrepreneurial companies to acquire or develop new technology, which they then market for themselves and their partner. In other cases, they collaborate with the smaller specialized companies to obtain fully developed technology, which they license for use in their product lines. There are many variations within this type of partnering, and the results have been very good for all involved, including fostering rapid growth of whole new industries based on biological and electronics technologies.

U.S. companies also are partnering with U.S. national laboratories. For many years, the U.S. Congress has been trying to improve technology transfer between these laboratories and industry. Success has only been apparent in the last five to eight years—since Congress recognized that laboratories and companies must work together on the development of technology, not just pass it over to companies for a royalty or other payment. Furthermore, the collaboration must advance the missions of all the parties involved. Today, national laboratories form partnerships with companies for high-risk research in support of their government missions. Companies are willing to share the costs of this research because they wish to use the resulting technology.

In some cases, individual companies form consortia to develop new products. One such collaboration involves the use of high-end super-computers in the design of new tire treads. Others efforts include the work of the semiconductor industry to develop advanced microchip manufacturing processes, collaborations to advance commercialization of high temperature superconductor devices, and exploration of the use of diamond films.

UNIQUE FORMS OF PARTNERING

Some of the more unique forms of partnering to develop high-risk technology are illustrated by Semetech, the Partnership for the Next Generation Vehicle (PNGV), and the U.S. Advanced Battery Consortium (USABC). A brief comment on each follows.

Semetech was formed by companies engaged in the manufacture and use of semiconductor devices. These companies were determined not to be shut out of the rapidly growing worldwide markets for semiconductors and related electronic products because of their inability to continue to develop and access state-of-the-art fabrication technology. This also was a concern of the U.S. Department of Defense. A partnership was formed between the government and these progressive companies to raise research funds and sponsor fabrication technology and equipment development that served the U.S. electronics industry. Initially the government and industry each contributed half of the $200 million in annual research funds, but because the partnership has been so successful, enabling U.S. companies to be in the forefront of semiconductor developments and to increase their market share in many key areas of electronics and computing, government support no longer is needed. Semetech was not directed to near-term market needs, but it ensured development of the tools to fabricate devices and systems at a reasonable cost and with specific capabilities. This partnership is an example of companies and the government working together to share the cost of pre-competitive technology development, the fruits of which the government and companies use to develop competitive offerings.

PNGV and USABC are somewhat different from Semetech. The U.S. government, in its role of preventing energy shortages and environmental damage, entered into partnerships with leaders in the U.S. automotive industry to fund high-risk research that would greatly decrease energy utilization of automobiles and reduce their environmental impact. The government provided half of each partnership's budget (tens of millions of dollars annually), but the automotive industry manages each partnership. The likelihood that current automotive technologies will be displaced by technologies developed under these partnerships should be maximized by day-to-day management by the industry that would use the new technologies. Both programs have been praised for their management approaches and technology advancements.[2]

LESSONS LEARNED

There are several lessons to be learned from the experiences of U.S. industry in attempting to improve technology commercialization, including the following:

- The best modern technology-based businesses succeed in today's marketplace through near-perfect communication and information flows among all branches of the enterprise. R&D personnel must be informed at all times of the business environment and must be included in planning and strategy development.
- All R&D activities do not have to be under the direct "control" of the business units they serve, but they, like R&D organizations within these units, must be tightly connected to the information flows.
- Universities receiving support by companies through grants and counsultanships must be closely informed about the companies' near- and long-term needs.
- Any R&D organization that is expected to serve a business must insist that all levels of the organization participate in near- and long-term business planning.
- Technology for eventual commercialization must be obtained from the best sources. World-class businesses view their R&D organizations as both sources for technology and advisors for technology acquisition.

---

[2] See "Lessons Learned Under the United States Advanced Battery Consortium," Abacus Technology Corporation, November 8, 1993, and the series of biannual evaluations by a National Research Council committee on the progress and management of the PNGV.

- World-class businesses understand and focus on their technology needs. They decide which technologies they should own exclusively and developed internally and which they may share and co-develop with others.
- The focus of R&D management always must be to access technology in the fastest, lowest-cost, and most effective way. Therefore, partnering with others for development of technology should be at the forefront of R&D management interests.
- Government can and should play a critical role in the development of certain new high-technology areas. It should not hesitate to partner with private industry when a vital industry has become too frail in the world marketplace for the best interests of the country, research serving the long-term interests of the country is too risky for any firm or combination of firms to perform, firms have better facilities and more experience than government laboratories to conduct the research, and the combined effort could result in lower costs and shorter time-frames for effective development.
- Government should consider allowing industry partners authority to manage joint efforts when the results are to be commercialized.
- In most cases, government partnerships with industry should involve groups or consortia of companies. Alternatively, provisions should be made to share the fruits of a partnership with all members of an industry without too much delay.
- Government national laboratories should be encouraged to work jointly with consortia and individual companies in high-risk areas when the research serves the mission of both parties.
- University relationships with companies should be encouraged to evolve toward *quid pro quo* partnerships while at the same time protecting the fundamental missions of universities to be a provider of basic research to society.

These lessons may be useful when considering ways to improve the connections between Russian universities and institutes and foreign firms and capital sources—connections that could help invigorate the Russian scientific establishment and improve its capability to restructure and rebuild technology-based industries. In the context of technology commercialization, the salient questions are:

- How can a significantly greater number of collaborative partnerships between Russian universities and institutes and industrial concerns in western countries be initiated?
- Can recent Russian immigrants living in the United States help initiate such partnerships without compromising their loyalty to their U.S. industry employers? (Many of the larger companies that have established research partnerships with Russian universities and institutes have effectively used their émigrés in these efforts).

- How can bilateral collaborations be established more quickly and with less misunderstanding?
- How can better understanding of western industrial business practices be imparted to Russian collaborators and government officials?
- How can Russian scientists and engineers more effectively contact Western firms that might benefit from collaborative relationships?

RESEARCH, TECHNOLOGY DEVELOPMENT, AND COMMERCIALIZATION

David McNelis
Research Triangle Institute

INTRODUCTION

U.S. research institutions and research universities receive support through grants from and contracts with the federal government. Just a short time ago, because of federal budgeting issues, many of these institutions, particularly the independent "for profit" and "nonprofit" entities and to a lesser extent the research universities, experienced a significant funding crisis. The crisis paled by comparison to that in which Russian institutions find themselves, but it provided sufficient incentive for U.S. research institutions to develop innovative strategies for obtaining long-term program funding.

The crisis was precipitated by an impasse in Congress over how to achieve a balanced budget. This impasse triggered the withholding of funding for much of the federal extramural research program. In general, no new research contracts were initiated for several months. In addition, certain agencies stopped work on programs that already had been approved and funded. Nearly every authoritative source indicated that government sponsorship of applied research would decline dramatically during the next several years, possibly a decade, and that support for fundamental research also would decline, although less precipitously. Many organizations with significant nonfederal sponsorship or with federal programs for which funding had not been withdrawn were able to shift their investigators into funded work. In many cases, however, staff members were furloughed and, in some cases, terminated. Many businesses failed after months of government inactivity.

The research funding crisis in the United States differs from that in Russia in two respects. First, the U.S. crisis was short-lived, whereas the current Russian crisis continues. Second, even during the U.S. crisis, the institutions and individuals conducting research, although fewer in number, were expected to be fully supported with state-of-the-art equipment, facilities, and resources. There is no such expectation in Russia today.

This paper addresses several positive changes resulting from the U.S. crisis, including some changes that made some independent R&D organizations even stronger and more competitive. The experiences and reactions of one organization, the Research Triangle Institute (RTI), are offered as a case study. It also incorporates observations made during the committee's visit to Moscow and St. Petersburg in November 1997. Although the paper extends beyond technology commercialization, it treats many of the issues that research organizations must address in this effort.

RTI is one of perhaps ten large-scale, basic and applied research, nonprofit organizations in the United States. A much larger group of smaller organizations generally follow the same policies and practices for business development and organizational management. Many of these policies and practices are dictated, at least in part, by the sponsors of research programs. Others have been developed over time in the interest of good business.

RTI was founded nearly 40 years ago by the three major universities in the local area—North Carolina State University, Duke University, and the University of North Carolina at Chapel Hill. It was created by those academic institutions and visionaries in the State of North Carolina to provide research opportunities for graduates and thereby retain the graduates as an intellectual resource for the region. Shortly after its creation, RTI became financially and administratively independent of the universities. In accordance with its charter, however, about one-half the members of its Board of Governors are associated with the universities.

RTI has approximately 1,500 investigators and staff members involved in basic and applied research and engineering. Most of the staff is located in Research Triangle Park, North Carolina; some researchers are located at RTI offices or facilities in Washington, D.C.; Hampton, Virginia; Cocoa Beach, Florida; and a number of small project offices in locations around the world.

RTI is organized as a not-for-profit corporation under a section of the U.S. Tax Code (501.C.3) that encompasses entities operating exclusively for religious, charitable, scientific, or educational purposes. As such, it is exempt from federal income taxes, with the exception of taxes on unrelated business income. Such income is earned when a university or a nonprofit research institution realizes gross income from any regularly conducted business that is not substantially related to the purposes cited above. It may be determined that questionable activities are not subject to unrelated business income tax; even if they are taxable activities, the revenue may be offset by associated expenses. State income taxes, intangible taxes, real property taxes, and sales taxes are governed by the state; and in North Carolina, these not-for-profit organizations are exempt from all such taxes. No part of the net earnings of a not-for-profit institution can be used to influence legislation or to participate in a political campaign on behalf of any candidate for public office.

R&D MANAGEMENT

The projected decline in funding by the federal government for research institutions caused the more proactive of these institutions to reassess their missions, strategic objectives, and management policies and practices. As an organization, RTI recast its mission statement to reflect this new market economy and examined four central management issues: strategic planning, cost/pricing, program development/marketing, and competitive intelligence. It retained, however, a business environment in which technical innovation could flourish and in which new ideas could be generated to attract investment.

Strategic Planning

RTI benefits from a strong reputation and facilities. However, the projected decline in federal funding forced RTI to create a new vision and redefine its strategic objectives as well as re-evaluate its markets in light of its capabilities and resources. Market shares were reviewed by government (department by department), commercial, state, and other sectors; by program; and by expertise. The federal budget was scrutinized for funding of new programs and program areas and for those marked for funding reductions. Market opportunities in the state and private sectors also were identified and assessed. Finally, market share projections were forecast on the basis of the best available information and estimates.

Costing/Pricing

RTI's costs, like any organization's costs, can be adjusted by modifying the components that comprise them. For example, RTI can reduce the number of administrative personnel supporting the organization or alter the mix of staff working on a project. Pricing, however, is the amount that a sponsor is willing to pay for a service regardless of cost. When working for the federal government, RTI must use cost-based pricing, meaning that it cannot charge more than its cost. When working for industry, however, RTI can use market-based pricing, meaning that market conditions control the price paid for such services. For example, RTI could, because of risk, include a contingency factor in its pricing for commercial work.

To be price-competitive, research organizations usually attempt to categorize their costs and rates. For example, they would price equipment- or facility-intensive projects differently than office-based tasks, wherein little or no special equipment is required. In this way, they can compete with others who may specialize in only one of these categories.

Program Development/Marketing

For a long time RTI depended extensively on personal contacts between RTI principal investigators and federal agency project officers to secure contracts. A review of program sponsorship indicated a lack of sufficient market diversity within the federal sector and a grossly under-represented commercial market share. To be more competitive, RTI recognized the need to be more proactive at all organizational levels in its marketing activities. Therefore, it used some overhead resources to hire senior marketing personnel and established sector marketing teams.

RTI's marketing strategy had to take into account the distinct differences between working for the federal government and working for the commercial sector. Commercial work typically presents a more difficult marketing challenge. Unlike the government procurements, which are publicly advertised, industry procurements often are only advertised via word of mouth, by personal contacts, or through industry representatives. In addition, industry requires work to be done quickly and, because of its concerns with intellectual property issues, confidentially.

Competitive Intelligence System

Every research and technology organization uses some system to obtain and analyze information about competitors. Competitive intelligence systems range from intuitive reasoning to sophisticated computer-based programs that incorporate searchable text and data files. The development of well-conceived competitive intelligence systems can serve organizations well. In addition, much like the creation of WWW home pages, developing such systems can be a marketable service.

INTELLECTUAL PROPERTY AND COMMERCIALIZATION

Unlike most nonprofit research institutions, which typically receive 70 to 85 percent of their funding from the federal government, RTI is 100 percent self-supporting and currently operates with an annual research budget of about $150 million. It is an example of a not-for-profit entity that must make a "profit" to survive.

Figure 1 illustrates government and industry investment in the development of technology by nonprofit organizations and universities. Government funding, the white area in Figure 1, supports basic research, applied research, and development through a variety of agencies and programs. Although typically not directly involved in the commercialization of any technologies resulting from these programs, the government during the past few year has developed new programs for partnering with industry to move demonstrated technologies

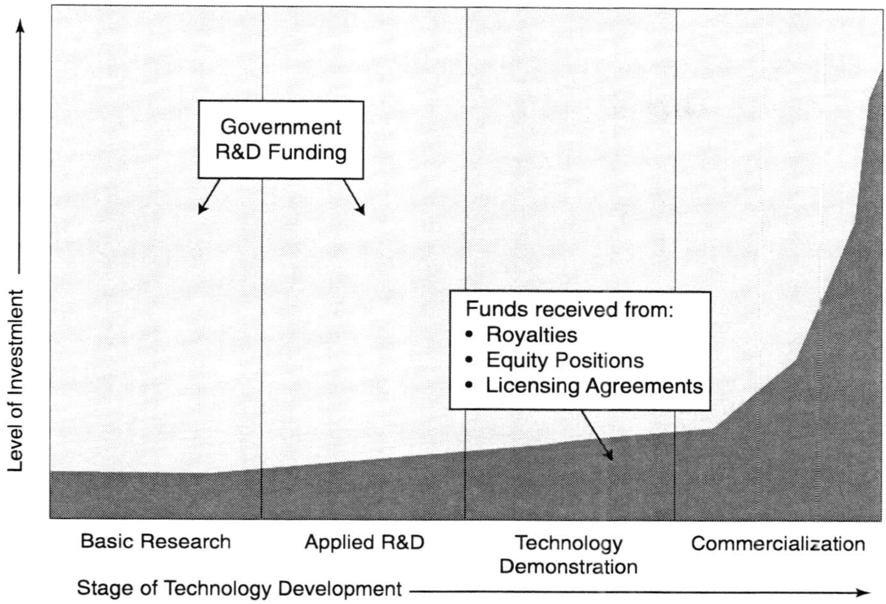

FIGURE 1 Government and Industry Share of R&D Funding at Different Stages of Technology Development

toward commercialization. Industry investment, the dark area of Figure 1, primarily supports the latter stages of the technology development cycle. As Figure 1 illustrates, government funding for technology development peaks at the research stage, whereas private sector spending is greatest at the commercialization stage.

Most research and academic institutions in the United States have become increasingly aware of the potential rewards associated with commercializing their intellectual property. Over the past few years, the majority of these institutions have developed an infrastructure to support technology commercialization, and some have sought external resources this activity. Institutions also created this infrastructure because they realized many investigators were either too close to their projects or too focused on the technical outcome of their research to recognize the commercial potential of their results. Institutions also have begun to educate their investigators about patent issues as well as to facilitate the administrative processes associated with disclosure and patent searches and with negotiation of licenses, equity positions, and royalties.

For some research institutions, income from the commercialization of intellectual property is a principal source of funding for the stimulation of new research efforts and for the continuation of work on projects through the "Valley

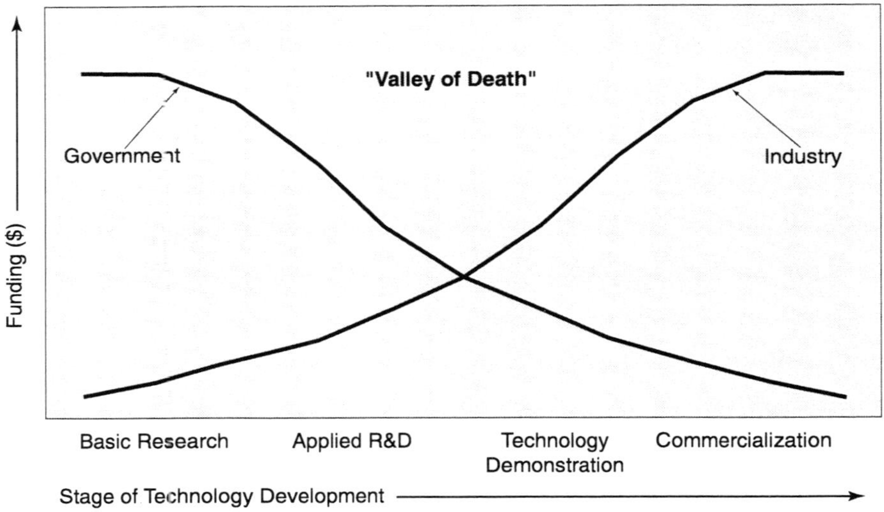

FIGURE 2 Valley of Death

of Death"—the stage of development at which technologies can die for lack of support. This is the period, as Figure 2 illustrates, when governmental sponsorship drops off and industry sponsorship has yet to be established. Patent royalties, equity positions in spin-off companies, and technology licensing agreements can fill the gap between these funding sources.

Unlike universities, which traditionally concentrated their research activities on basic or fundamental research, independent research institutions generally have focused on applied R&D. One of RTI's units has the mission of assessing and marketing technologies developed by government and commercial laboratories. The unit analyzes the size and stability of a technology's potential market, the presence and strength of competitive technologies, and any laws or regulations that could affect the technology or its use. Leveraging its expertise and technical network, RTI also identifies sources of new technology that could be imported into an industry to improve cost savings, market share, or manufacturing processes. For industrial clients, RTI facilitates the licensing of corporate technology, identifies potential partners, negotiates with possible licensees, and otherwise assists in establishing R&D collaborations.

For government laboratories, RTI provides a range of services, including identifying promising technologies and assessing their commercial potential, promoting technology to industry, developing cooperative research and development agreements between the federal laboratories and third-party organizations, performing market assessments, and providing business development assistance. For example, RTI coordinates a wide variety of licensing opportunities for NASA centers at the Langley, Kennedy, Ames, and

Goddard locations—from transferring the methodology for converting nitrogen oxides waste to fertilizer to seeking partners for commercializing the Smart Surgical Probe, a system that evolved from telemedicine research on astronauts. An RTI-government partnership also commercialized a hand-held imager developed by NASA's John C. Stennis Space Center. The device is capable of imaging the invisible flames of alcohol and hydrogen fires during the daytime, as well as penetrating moderate smoke, fog, and mist. (NASA uses huge volumes of liquid hydrogen in flight certification, testing, and launches and, due to the risk of hydrogen fires, needed a low-cost device that could not just detect a fire but could indicate its precise location and extent.) After assessing the commercial potential of the device, RTI held an industry briefing to introduce fire fighting and safety companies to the technology and to invite them to submit commercialization plans. NASA selected one company to receive the exclusive patent license rights to the Stennis fire imaging technology. The device has now been established as the first affordable commercial product for imaging invisible (ashless) fires and hot embers.

LESSONS LEARNED

RTI's positive and decisive actions have allowed it to prosper across essentially every organizational element of the institution despite the sharp decline in new programs and funding availability from the federal government. The steps outlined above could improve the competitive posture of any organization. Russian research institutions therefore should give serious consideration to the interrelated issues of strategic planning, management and staff training, and marketing in addition to the issues surrounding taxation and intellectual property rights. Also the opportunities for sharing government facilities and equipment seems important. The Russian government has huge investments in research instruments, infrastructure, and facilities which could be shared with independent research institutions and the private sector. Although barriers exist in the United States, government agencies do share capital investments in expensive equipment and facilities with universities and nonprofit research institutions. They also make excess and surplus property available to the private sector at a fair market price.

Strategic Planning

All too often, strategic planning is conducted only periodically, and then it is essentially ignored until the next planning cycle. To be effective, strategic planning must be continuous, involve all staff levels, and address all aspects of the business, including administration, competitiveness, resources, capabilities, and market development.

Management Training

Independent research organizations must not focus only on income and percent of products sold, but also on management and training. RTI has earned large returns on its extensive investments in managers in writing business plans, strategic planning, and the development of marketing plans. Every organization should carefully assess their training needs and the potential return on investment associated with well designed training programs.

Marketing

Over the last few years, the commercial use of the Internet and the World Wide Web has increased dramatically. Home pages, creatively constructed and strategically linked, can serve to inexpensively advertise and market products and services. The Web also can provide a wealth of information on markets and competitors. Of course, a competitor for one program or market could well be a partner for the next. And other new uses are emerging. For example, in early 1998 a university announced over the Internet a service for businesses impacted by reductions in the U.S. Department of Defense's budget. The university's Entrepreneurship Center offered to help these companies identify new business opportunities, diversify existing business bases, and facilitate the transition of products and services from defense customers into civilian government, commercial, and foreign markets.

CLOSING REMARKS

As an environmental scientist, I must close with a caveat. We are living in an industrial age. Since the industrial revolution, we have been encouraged to be entrepreneurs—to use our planet's resources to provide for our needs. Even today, governmental policies in most countries tend to favor exploitation. Compliance merely means to be as bad as the law allows. Sustainability, however, requires that we also allow for the needs of our progeny. The exploitation and alteration of our planet earth in an in credibly short amount of time is well documented and understood. Over the long haul, the economy is a wholly owned subsidiary of the environment, we must make the effort to care for our planet and its life forms in order for it to provide for us. The hope, therefore, is that each of us is mindful environmental sustainability in our technology development and commercialization endeavors. Otherwise, economic sustainability cannot occur.

VIEW FROM A NATIONAL LABORATORY

Alvin W. Trivelpiece
Oak Ridge National Laboratory

Approximately 600 "national laboratories," most with an annual budget of less than $50 million, are sponsored by U.S. government agencies. Of these laboratories, about 20 are national laboratories of the Department of Energy (DOE). Ten of these are major, multiprogram laboratories with annual budgets in excess of $100 million:

Argonne National Laboratory (ANL)
Brookhaven National Laboratory (BNL)
Idaho National Energy and Environmental Laboratory (INEEL)
Lawrence Berkeley National Laboratory (LBNL)
Lawrence Livermore National Laboratory (LLNL)
Los Alamos National Laboratory (LANL)
National Renewable Energy Laboratory (NREL)
Oak Ridge National Laboratory (ORNL)
Pacific Northwest National Laboratory (PNNL)
Sandia National Laboratory (SNL)

Each of these laboratories carries out different aspects of the DOE's mission. As a result, they have different capabilities and facilities. Some are involved in work related to national security and nonproliferation, some in basic and applied research and development in various areas of science and technology, some in energy production and supply, and some in environmental remediation and waste management. In many areas of activity, no clean line of demarcation separates one laboratory from another.

Twenty years ago none of these laboratories had extensive contact with business or industry from the point of view of technology transfer. Today, that has all changed. The change did not occur suddenly, although some events have accelerated the commercialization of science and technology developed by the laboratories. Most notable was the discovery of high-temperature superconductivity and the concern that the United States might not have a sufficient opportunity to rapidly develop products or services based on this discovery. This concern led to the establishment of Superconducting Pilot Centers at ANL, LANL, and ORNL. These Centers permitted, for the first time, joint industry-government funding of research and development whereby the industry partner would have proprietary rights to intellectual property that might result from the project. This approach was productive, and after a few years, legislation was passed permitting this approach for many other such arrangements. This 1989 legislation created the basis for Cooperative Research and Development Agreements (CRADAs). Today, the Department of Energy's national laboratories are engaged in more than 1,500 CRADAs supported by more than $200 million of DOE funding and a comparable level of funding by the industry partners.

CRADAs are not the only means of technology transfer from DOE's national laboratories. Industry can contract for work performed at the laboratories when the work does not compete with other activities of private industry. Typically, such work involves the use of unique scientific facilities at a laboratory—for instance, high resolution electron microscopes, synchrotron light sources, or research reactors—to analyze some material. Scientists and engineers at national laboratories can, under certain circumstances, serve as consultants to industrial organizations. In some cases, employees of the laboratories are able to gain control of the intellectual property they have helped create and start a company that takes advantage of that property.

Russia has laboratories that are in many respects the equivalent of national laboratories in the United States. Some of these laboratories had been dedicated to national security pursuits that are no longer a critical imperative. The question is how Russian laboratories can benefit Russian industry and the economy. Technology transfer may not be of major direct benefit in the short term; however, forging links between Russian laboratories and industry likely will be of great benefit to the Russian economy in the longer term. To illustrate the potential, in the United States it is estimated that over 50 percent of the jobs created in the past ten years are related to science and technology developed more than a decade ago by government, industry, and academia. For such benefits to be realized in Russia, the appropriate legal and economic framework, together with a system of personal incentives to stimulate invention and its exploitation, must be in place.

One key to stimulating technology transfer at DOE national laboratories has been the establishment of a system whereby the inventor is rewarded for a patent and shares in the licensing fees, the institution shares in the proceeds, and the federal government realizes some direct benefits from the fees. This system has been in place for about 10 years. At the Oak Ridge National Laboratory, the fees earned for licenses amount to about $1million per year. Some of this money is used to offset the costs of filing and defending patents and some of it goes to the inventors. This reward system has spurred the creation of intellectual property and its dissemination to industry.

As Russia sorts through its near-term problems of the economy and overcapacity of manufacturing, it should also devote attention to the establishment of a system that rewards invention and the development of intellectual property. Such a system should be tailored to meet the requirements of the Russian social and economic structure. In the United States, the laws pertaining to intellectual property rights and inventions are intended to maximize "fairness of opportunity" and minimize "conflict of interest." Therefore, for example, a director of a laboratory cannot approve a CRADA for a company of which he or she is part owner. It is not possible to anticipate all the circumstances that might lead to such problems in the United States or in Russia; however, it is important to develop a Russian system that Russians trust. Failure to do so would lead to a very weak system.

Another key to the stimulation of technology transfer activities at DOE national laboratories has been increasing the number of contacts between scientists and engineers at the laboratories and their counterparts at industrial sites. One way such contacts have been increased is through the "User Facilities" operated at ONRL and other DOE national laboratories. These facilities were created for various purposes. For example, the High Flux Isotope Reactor (HFIR) was intended to produce various isotopes when it was built in 1966. It also was designed to permit experiments on neutron scattering and diffraction. Today, scientists and engineers from around the world use HFIR for neutron scattering experiments to study the structure of materials and the stresses in materials caused by welding or other processing methods. There is a two-year backlog of experiments that various academic and industrial organizations want to perform at DOE user facilities. To handle some of these experiments, a $1.3 billion Spallation Neutron Source (SNS) is planned at ORNL. The SNS accommodate more than 1000 users per year, and it will be operational in 2005.

DOE user facilities generally are made available to scientists and engineers without charge, unless the user wants to have proprietary rights to the data they produce at the facility. In these cases, the cost of using the facility is charged to the user. These user facilities bring many visitors into the laboratories, and as a result, new ideas are generated that may lead to inventions. Perhaps the most important benefit comes from the contacts that take place between visiting scientists and engineers and the host laboratory staff.

The structure and function of the system of laboratories in Russia are quite different from that in the United States. Even so, Russian laboratories might find ways in which facilities such as accelerators, reactors, and computers can be used by scientists and engineers from outside the host institution to develop new products or services.

THE ROLE OF INDUSTRIAL INSTITUTES IN CREATING AND MAINTAINING RUSSIA'S INDUSTRIAL POTENTIAL

S.S. Ivanchev
NPO Plastpolimer

ORGANIZATIONAL STRUCTURE OF SCIENCE IN RUSSIA

In the post-war period, only two national scientific structures—the American and the Soviet—have been able to conduct research over the entire scientific-technical spectrum and to advance substantially in science and technology. In the USSR, this was achieved by the high priority that the government placed on science.

In the USSR, scientific-research units were structured along two lines. First, institutes of the USSR Academy of Sciences and the Academies of Sciences of the union republics carried out basic research in various fields of science in accordance with each institute's profile. A second group of institutes—industrial institutes—conducted technological development work for their corresponding industrial branches and were responsible for the development of these sectors. In many cases, industrial institutes were granted rights as lead research units in given technological fields. Duplication of science and technology topics by the various institutes was not permitted. Both the basic science and industrial institutes were state organizations; however, they were directed by different branches of the government. The first group of institutes was financed through the Academy of Sciences system, while the second group was supported through the appropriate ministries.

The industrial institutes played a defining role in creating new technologies for all branches of industry. Although the Academy institutes elucidated new scientific principles and natural laws facilitating the creation of new industrial technologies, the industrial institutes bore responsibility for the actual development of these technologies. In fact industrial institutes carried out all design work, right down to the level of organizing production and providing the scientific documentation for the operation of the production facilities.

The industrial institutes represent a complex of technologically-oriented scientific-research organizations. The basic task of these institutes is to conduct scientific research, engineering, testing, and design work aimed at the creation of industrial facilities, installations, and technologies and to provide scientific documentation for the work of these facilities and installations. In the chemical industry there were about forty such institutes, working in contact with institutes of the Russian Academy of Sciences (RAS) and institutions of higher education.

In the USSR, the profile of each industrial institute was determined on the basis of the product principle and was connected with the development of materials closely related by the technological characteristics of their production (e.g., polymerization plastics, polycondensation resins, polyurethanes, poly-epoxies, phenolformaldehyde resins, polyacrylates and polyvinyl chloride, and plastics reprocessing). All of these institutes, which were the only ones in the country carrying out their particular type of work, were assigned lead functions and responsibility for the development and application of these materials For example, the lead institute for polymerization plastics (now the Plastpolimer Joint Stock Company Institute) was and is responsible for the development and implementation of technologies for producing four types of polymerization plastics: fluoropolymers, polyolefins, polystyrenes, and polyvinylacetate plastics. For the 52 years of its existence, the institute developed and implemented technologies for a broad assortment of fluoropolymers for the domestic market, including for the aerospace, aviation, and defense industries. The institute has helped to ensure that domestic demand for fluoropolymers in the Soviet Union and now Russia has been fully met, making the exportation of such polymers a possibility.

The institute has developed several notable technologies. For example, the institute developed a technology for producing thin layers of polyethylene in a tubular reactor. Five production units have been built on the basis of this innovation. It also developed a technology for producing high-impact polystyrene and another for producing acrylonitrile/butadiene/styrene (ABS) plastics by mass polymerization, both of which have been put into production at several plants.

The institute has made substantial improvements in production processes created on the basis of technologies purchased from foreign firms. For example, the institute worked out a domestic technology for the dimerization of ethylene in butene and created two production units for obtaining butene as a comonomer in the production of linear polyethylene by the gas-phase method using equipment purchased from the Union Carbide company. For its development work, the institute used a test facility featuring an entire arsenal of experimental equipment. It has used the equipment in such processes as the polymerization of olefins (including under high pressure), fluoromonomers, styrene, and vinylacetate and in the creation of polymer composites.

The plastics institute has always worked closely with RAS institutes, including the Institute of High-Molecular Compounds, the Institute of Chemical Physics, the Institute of Petrochemical Synthesis, and the Siberian Branch Institute of Catalysis, and with higher education institutions, including St.

Petersburg University, the Technological Institute, Moscow State University, and Kazan Chemical Engineering University.

CURRENT CONDITION OF THE INDUSTRIAL INSTITUTES

Beginning in the 1990s, the situation was substantially altered for scientific institutes. The position of industrial research and development changed in the wake of the breakup of the USSR, various economic reforms (including a change in the management of production and the liquidation of Gosplan [the USSR State Planning Committee]), a change in the responsibility of ministries and the subsequent elimination of several ministries, the granting of independence to industrial enterprises, and the privatization of organizations and enterprises as well as strategic and tactical mistakes during the transition to a market economy. Under their new organizational structure, privatized industrial institutes were almost totally deprived of state financing. State affiliations were maintained only by Academy of Sciences institutes, several large scientific institutions which had performed most of their work for the defense complex before receiving the status of Russian Scientific Centers (a total of 56, of which 5 are in the field of chemistry), and a few industrial institutes. The industrial institutes were supposed to obtain work and finances through contracts with enterprises in their particular branches of industry.

In the first year, these structural changes seemed acceptable. Freed from state guardianship, factories had an interest in receiving help from the industrial institutes in updating their technical documentation, retooling their facilities, analyzing marketing data, and creating new production capacities. However, the political and economic changes soon began to have a significant impact on the financial and economic condition of industrial enterprises. A general decline in production began in practically all industries, including the chemical industry; this resulted in a fall in plant capacity utilization, widespread nonpayment for products, and other well-known features of the crisis.

The changes in political boundaries also contributed to the collapse of production. The breakup of the USSR resulted in the drawing of new borders, geographically isolating raw material suppliers from processing plants, the introduction of customs limitations, increases in the cost of energy and transportation, and the opening of the market to major foreign firms. All of these events, for which producers were unprepared, were detrimental to the work of factories and hindered the sale of their products. In the chemical industry, the collapse in production also was brought about by the orientation of the chemical industry towards the military-industrial complex, the tradition of using relatively cheap raw materials and human resources, the unjustifiably isolated locations of certain facilities, and centralized planning and distribution.

Together these factors led to a sharp reduction in the production of goods and, consequently, to an extremely low rate of profitability. In some cases, factories and plants went bankrupt. In the chemical industry, which produces about 7 percent of all industrial output, employs 6 percent of the industrial-manufacturing personnel, and possesses about 10 percent of the total value of Russian assets, the level of production in 1996 was less than half of the 1990 level. The efficiency of many factories had fallen both in terms of energy consumption and specific expenditures in the cost structure. The proportion of worn-out equipment rose sharply, and profits fell to a third of the previous level. Many enterprises ended 1996 with losses. Utilization of production capacity fell to an average of 36 percent between 1991 and 1996, which naturally led to an increase in production cost and a reduction in market competitiveness. Finally, the lack of preparedness to respond to new financial and tax policies led to a shortage or complete absence of working funds at enterprises and a halt to research and development activities.

Having no profits, factories could provide their industrial institutes with neither work nor financing. Given shortages in the funding of science as a whole and sectoral science in particular, industrial institutes found themselves in an extremely critical situation. Left without financing, the institutes began to cut personnel and lose test bases and facilities created with great effort. They were unable to update their equipment and instruments. Some of them ceased to exist or are on the verge of collapse.

STABILIZATION MEASURES UNDERTAKEN BY THE RUSSIAN GOVERNMENT

Issues involving the stewardship of science are now handled by the Ministry for Science and Technology, which was created to replace the Soviet-era State Committee on Science and Technology. The ministry prepares the appropriate government decisions on science and creates legislative proposals aimed at maintaining and encouraging scientific activity and protecting intellectual property. While basic science is given priority in the activities of the ministry, industrial institutes are also the focus of the ministry's concerns.

The ministry works actively to improve the legal and economic climate for increased investment in science and technology. For example, it has modified financial and tax policies and created additional sources of financing for scientific-research organizations. The Ministry of Science also conducts a great deal of systematic work to help industrial institutes adapt to market conditions.

The ministry has granted the status of State Scientific Center of the Russian Federation to a number of large institutes that work in fields of great importance to the state. This status provides them with financing, significant tax advantages, and organizational assistance in restructuring themselves and attracting foreign and domestic investors.

Working with the Russian State Committee on Standards, the Ministry of Science has done much to create a system of certification for chemical products and accreditation for chemical plants. With the involvement of the State Scientific Centers, the ministry has established special organizations to improve the innovation activities of the industrial institutes. A great effort has begun on the accreditation of Russian scientific organizations. It has created industry associations to consolidate resources for carrying out complex projects—an effective means of attracting funds from enterprises to finance research. An example of such an organization is the Elastomers Association, whose members are producers of paint and varnish products.

The ministry actively helps sectoral institutes to cooperate with foundations which might finance their work, including the Russian Foundation for Technological Development, the Chemistry for the Ecology Foundation, and the International Science and Technology Center.

The ministry has introduced multi-channel financing through foundations accessible to any institute at the appropriate level of work. These organizations include the Russian Foundation for Technological Development, the Priority Research Foundation, and international foundations, such as the Soros Foundation and the International Science and Technology Center. It also has encouraged the participation of industrial institutes in international projects.

By a special decree, nongovernmental scientific organizations have received equal rights to benefits granted by the state to facilitate the activities of scientific organizations.

MEASURES UNDERTAKEN BY INDUSTRIAL INSTITUTES

There is a saying in Russia that saving a drowning man is a job for the drowning man himself. This saying underscores the idea that the solution to the crisis in which the industrial institutes find themselves will be determined in part by the enterprising steps that the institutes themselves take. Knowing their own characteristics, the nature of their work, and the scientific potential they have amassed over decades of work, the institutes often find interesting options for emerging from or easing the crisis. These options might be relevant only to a given institute, but sometimes they may be appropriate for other institutes as well.

Consider the solution that the Research Institute of Polymerization Plastics (Plastpolimer Joint Stock Company, St. Petersburg) found to secure work orders after the cut-back of state financing and the termination of investments in the plastics sector. After reviewing its technological research priorities, the institute focused its efforts on developing new projects to improve technologies for producing plastics and polymer composites with enhanced properties or better economic characteristics. Given the temporary difficulties with investments in Russia, the institute proposed many projects to foreign firms for continuation on a contract basis. Leading firms from Europe, the United States, and now Asia have signed contracts with the institute to carry out these projects. In addition, the institute offered testing services to be carried out in accordance with the scientific programs of these foreign firms. Since the beginning in 1993, the offering of testing services and of scientific projects for joint continuation have enabled the institute to carry out $300,000 to $500,000 of work per year on a contract basis—at least one quarter of its annual volume of work.

In just the past four years, the institute has initiated long-term contracts with such firms as 3M (U.S.), Dow Chemical (U.S.), Borealis, Neste Chemical (Finland), Akzo-Nobel and D M (Holland), and RSD (Austria). These contracts have enabled the institute, at least in part, to finance and maintain its scientific potential and to replenish its supply of instruments and equipment for research and technological work. In many cases, the terms of the contracts made it possible for the institute to retain the possibility of using the technologies it develops for other firms in its own enterprise.

In recent years the institute has been able to license sales of fully developed technologies. It perfected several technologies by improving certain domestically invented stages in plastics production processes. For example, as a result of skillful selection of efficient new initiating systems at many plants, the institute has been able to increase the quality and assortment of thin polyethylene while simultaneously enhancing the economic efficiency of production. The institute also has begun to improve the technology for olefin polymerization by using new catalytic systems based on metallocenes.

In addition to taking steps to secure contracts with foreign firms and profit from licensing sales, the institute has made other efforts to bolster its scientific potential. It has maintained the scientific council for granting academic degrees and its institute for training personnel at the graduate level. This effort to produce highly-skilled scientific workers also will enhance the institute's long-term viability.

Together, these steps suggest that the institute will be able to overcome the crisis situation, positively influence the activities of the factories in its industrial sector, and ensure the competitiveness of domestically-manufactured products on the free market.

What basic lessons have we drawn from the transition to a market system? First, industrial institutes should proceed with only those technological developments for which there likely will be a demand under market conditions. Second, they must carry out research and development work in a timely manner. If the time frame from project planning to commercialization is too long, a technological development can lose its novelty and, consequently, the interest of the market.

PROBLEMS OF TAXATION AND TECHNOLOGY COMMERCIALIZATION IN RUSSIA

Yu. O. Lebedev
Ministry of Science and Technology of the Russian Federation

Economic stabilization and growth in the gross domestic product depend on the effective use of the results of scientific-technical activity. Otherwise, it is impossible to raise the competitiveness of Russian goods, labor, and services; improve technological development; and attract investment resources for industry.

In Russia today, legislation defines the legal status of participants engaged in economic activity, the basis for creating intellectual property, and procedures for its sale or transfer. The concept of intellectual property implies that individuals and/or legal entities have exclusive rights to the results of their intellectual activity. These rights may include rights of authorship, priority in the securing of legal rights, and/or exclusive rights to the use of results.

As a rule, the most important intellectual property items in terms of socio-economic development are results stemming from scientific-technical work. The potentially high market value of such items and the possibility that their use would stimulate scientific-technical progress depend on (1) development of legal norms for protecting intellectual property from unauthorized use, and (2) formulation of government regulation on the use of intellectual property created with funding from the state budget. Protection of intellectual property from unauthorized use requires the development and adoption by the government of legal norms allowing physical persons and legal entities in both the state and private sectors to create, buy, and sell intellectual property while observing relevant property rights. Regulation of the use of intellectual property created with state funding means that the government not only controls the transfer of such property, but also develops conditions for its transfer. The government must ensure legal protection for intellectual property rights and create the infrastructure for their use, bring intellectual property to bear in the national economy to the maximum extent possible, license its intellectual property, and observe the rules of honest competition and limit monopolistic activities.

The legislative basis for the protection and defense of intellectual property can be found in the civil and criminal codes, the patent law, the law "On Copyrights and Associated Rights," and other special laws. These laws protect the interests of intellectual property owners both within Russia and in foreign markets. However, most of the legal framework was created and adopted in 1992 and 1993, and therefore it does not reflect the rapid institutional changes, particularly with regard to property, that have taken place in Russia. Certain imperfections in the laws, as well as the absence of an adequately developed system of judicial redress concerning intellectual property rights, may be compensated for, but only to a certain extent, by careful preparation of contracts for the performance of scientific-technical work.

Unfortunately, in practice the government has declined to control the use of scientific-technical results obtained using government funds. The consequence had been the spontaneous redistribution of rights to such results, the ineffective use of research results, the development of many undefined and contentious relationships, and the violation of the legal rights of patent and copyright owners. These developments highlight the need to examine the entire set of issues associated with the creation, legal protection, and introduction into the economy of the results of scientific-technical activity and to formulate the basic position of the government on the questions of intellectual property arising in the scientific-technical sphere. These issues include:

- The government should encourage commercial sales of scientific-technical results, with appropriate legal protection, to stimulate profitable activities on the part of scientific-technical organizations.
- The government should not permit industrial enterprises to appropriate research results from scientific-technical organizations for the so-called "interests of the industrial sector." Instead, the government should require enterprises and research organization to develop legal, contractual relationships, primarily through licenses that provide economic incentives for scientific-technical work.
- The government should require that scientific-technical work funded by the federal budget is carried out on the basis of legal contracts and associated arrangements. These contracts should stipulate the rights of (a) the creators of the research results, (b) the scientific-technical organizations that employ these individuals, and (c) the state as represented by federal executive agencies or their officials. Demarcation of these rights must be tied to specific obligations and must ensure legal protection of intellectual property and its future use, especially its commercial use.
- The government should use the principle of competition to decide financial and other types of support to organizations and entrepreneurs in the scientific-technical sphere. It should make such support available on a repayable or non-repayable basis. In the process of allocating such support, the government must give priority to the protection of intellectual property,

the efficient use of which will facilitate the satisfaction of its own and society's most pressing needs.

- Given current economic and budgetary conditions, the government should not create state organizations to commercialize intellectual property, including property created using federal funds. Instead, it should identify a range of organizations with successful experience in commercialization, work with these organizations on the basis of agency agreements, and subsequently recommend their activities to the scientific-technical community. The government should not place all opportunities for the commercialization of intellectual property in the hands of a limited number of organizations, even organizations representing state interests. Nor should it create a single method for calculating the value of intellectual property; questions about the value of intellectual property should be resolved by organizations that handle the commercialization of intellectual property.
- The President and Government of the Russian Federation should encourage judicial organs to increase its protection of owners of intellectual property in the scientific-technical sphere against the property's unauthorized use, including unauthorized use by the government. Such efforts could begin with an analysis by the high courts of cases and rulings in the criminal, arbitration, and administrative systems with regard to such violations. Recommendations for protection of intellectual property then could be developed for courts at the local as well as the state level.

The establishment of clear regulatory mechanisms for protecting intellectual property will promote the effective, fair, beneficial, and broad use of scientific-technical results (from publication of research to production of technology-intensive goods and services). These proposed measures will be only the first stage of a large and very important effort to stimulate the market for intellectual property in Russia and provide highly effective means for bringing scientific-technical achievements to bear in the national economy. However, these measures must be adequately financed and receive other necessary government support. Given that the government has only 18 percent of the country's financial resources at its disposal, state tax and amortization policy also begins to play a very important role. The draft Tax Code of the Russian Federation, which is currently being considered by the parliament, sets forth the new tax and amortization policy.

The Russian tax system includes federal, regional, and local taxes. Individuals are subject to taxes, as are legal entities, including scientific-technical organizations. Specifically, the draft Tax Code classifies the following as taxable: any scientific-technical work, services in the science sector, and any payments received as compensation for the use of or granting of authorship rights to any scientific project or as reward for information related to scientific experience. A new and important measure proposed in the draft code is the opportunity for an organization to receive an investment tax credit for carrying out scientific-technical work.

Some tax exemptions are allowed for scientific-technical organizations. One of the basic exemptions involves the value-added tax on goods, labor, and services produced in Russia and in other countries in the Commonwealth of Independent States (CIS) and on goods imported into Russia. The only goods exempted from this tax are those brought into the customs territory of Russia as non-reimbursable technical aid for joint scientific-research efforts carried out under contracts with foreign educational and scientific organizations. Such goods also receive preferential customs tariff treatment. The absence of other exemptions to the value-added tax brings scientific-technical work into conformity with the norms of the Civil Code of the Russian Federation, which categorizes contracted scientific-technical work as profit-making or entrepreneurial activity.

Profits (income) of organizations are also subject to a federal tax. The draft code establishes the possibility for an organization to deduct expenditures for scientific-technical work from its income tax base. Included are expenditures for creating new or improved goods, labor, and services. These expenditures may be considered as material costs or may be amortized. The draft code also creates a tax deduction for humanitarian activity, which includes transfers of funds to government and municipal organizations for scientific activities.

The draft Code would exempt certain kinds of income of individuals. For example, the following would not be taxed: state stipends for graduate students, stipends provided by organizations, grants (nonrepayable aid) provided for the support of science by international or foreign organizations, and noncommercial or charitable funds registered with the Ministry of Science and Technology. Although they pay income tax, authors and inventors of scientific works can deduct related expenses that total 20 percent or more of the total income they receive from the use of their intellectual property rights.

Several other exemptions from specific federal taxes should be noted. In particular, individuals or organizations which have received parcels of land for use as scientific test sites are exempted from paying land tax. Also exempt from this federal tax are scientific organizations that collect samples of animal life and aquatic biological resources for scientific purposes. Likewise, scientific-technical activities conducted in connection with forestry studies are exempt from the federal forest tax. Finally, the federal water tax does not apply to the conduct of state scientific monitoring of water and other natural resources.

Of the regional taxes, the tax on the property of organizations is of the most interest. State institutions financed from the federal budget, along with other scientific-technical organizations accredited by the state in accordance with requirements set forth in the federal law "On Science and Governmental Scientific-Technical Policy," are exempt from regional taxes on property used in research and experimental production. These requirements are the following: scientific-technical activity is one of the basic activities of the organization, a scientific-technical council operates within the organization, and at least 70 percent of the income of the organization comes from scientific-technical work.

It is apparent that the new tax policy should promote the reform of Russian science. First, the policy would compel scientific-technical organizations to cease activities not relevant to their missions, dispose of excess property and land not used in carrying out this work, and concentrate their attention on scientific-technical efforts. Furthermore, the policy would encourage scientific-technical organizations to avoid contracts that do not ensure the appropriate financing and fulfillment of obligations by customers and to seek only those contracts oriented toward creating scientific-technical products that can be quickly and profitably commercialized. At the same time, organizations will have new incentives to take all necessary measures for legally protecting the intellectual property they have created. In this regard, change in the system of payment of workers in these organizations is needed to increase material incentives for inventors and product designers. In addition, development of pragmatic business relations among participants in scientific-technical activity also is needed. Finally, the new tax policy would give significant preferences to international scientific-technical cooperation, which should facilitate intensification of such cooperation.

All of the enumerated elements of internal organizational restructuring will consolidate an important trend that already is becoming evident in the reform of Russian science. Only those scientific-technical organizations with a high-level commitment to commercialize the results of their work will be successful. Such commitment should take into account that the value of commercialized scientific-technical results will grow. Significant non-budgetary sources of financing may become available for scientific-technical activity, allowing scientific-technical organizations to become more independent.

As a result of these developments, some changes in the functions of government will be necessary. In the future, the government should have the following basic roles: collection of information about the research results of scientific-technical organizations, analysis of this information and identification of basic problems, adoption of policies necessary for the activities of scientific-technical organizations, and monitoring the implementation of these policies. Such a course will require increased activism on both the part of representatives of the scientific-technical community and among governmental administrative agencies. In the near term, there is no other realistic path to follow.

COMMERCIALIZATION OF SCIENTIFIC AND TECHNICAL DEVELOPMENTS AT HIGHER EDUCATION INSTITUTES

V.S. Litvinenko
St. Petersburg State Mining Institute (Technical University)

Founded in 1773, the St. Petersburg State Mining Institute (Technical University) was the first higher technical educational institute in Russia The institute has played a key role in the development of the geological sciences and related industries. The only mining university of a polytechnic type in the country, it covers all aspects of mineral exploration and assessment of metallurgy. The institute traditionally has had close connections with production, and it currently has more than 100 professors and 600 candidates of science working actively in the fields of geology, mining engineering, and metallurgy.

As a leading higher education institution, the institute plans its activity with following considerations in mind:

- social reforms, structural reforms of science and industry, development of educational services;
- reforms in the professional education system and transition of education to a multilevel structure; and
- changes in the status of many higher educational institutes and the transformation of some into academies and universities.

The institute's capability to respond to the changes guarantees its survival and further development.

PRIORITY TASKS OF THE UNIVERSITY

Three years ago, the institute's senate determined that the primary activities had to include scientific research. This research must be conducted on the basis of creation a flexible management structure. It must encompass scientific developments related the needs of society and the state and effectively apply scientific potential to improve the quality of specialists. The specific tasks include:

- retraining the administrative staff;
- improving the managerial structure of the university;
- developing norms for increasing the responsibility and discipline of the employees at all levels;
- creating a system for forecasting, planning, controlling, registering, and analyzing condition related to managerial and financial activity;
- creating a network of programs for quick and comprehensive problem solving at all levels;
- applying innovative decision-making methods;
- providing multiple financing sources for all university activities; and
- improving the efficiency of resource use.

Given the country's economic problems, the institute has to address many issues to create and maintain an appropriate environment for work while improving the quality of scientific and educational activities.

DEMAND FOR UNIVERSITY SCIENCE

An analysis of the demand for institute services revealed that the following factors have affected the basic production assets of industry: low quality of domestic industrial products, active competition from imported products, and a steep decline in production and investment. To identify customers for university research, the institute also analyzed its sources of research funding. The effort revealed the following distribution:

- funds from enterprises: 87 percent
- state budget funds: 4.5 percent
- foreign funds: 5.0 percent
- off-budget funds: 2.0 percent
- other sources: 1.5 percent

These investigations suggested that:

- university science can be of interest to domestic enterprises,
- scientific work has to have a short payback period,
- introduction of a new product does not demand a large startup investment (up to $5 million), and
- enterprises are less concerned about securing patent rights than about securing their rights to use the final products.

On the basis of these insights, the institute identified requirements for its scientific research.

Structural changes

The institute's laboratories were merged according to the industries they served (oil, gas, refinery, etc.). Innovation and technological laboratories were created for each area. A new laboratory—an analytical, innovation, and technological center—was established to:

- create databases on productive assets,
- amass geoinformation,
- solve technological problems of production enterprises,
- assess industrial infrastructure conditions, and
- create databases for regional innovation programs.

Personnel policy

To attract and retain the best scientists for research, the institute formulated new personnel policies. All research engineers were subject to unannounced appraisals. As a result, the best professional scientists were identified and subsequently retained during staff reductions. Labor contracts were negotiated with all employees. And a differential system of remuneration of labor was introduced.

Research Areas

The institute has undertaken important applied research in many areas.

Geology

More than 22,000 fields of various minerals are found in Russia. Many useful minerals have been salvaged from depleted mines, the refuse of mills, and the waste products of metal manufacturers. These sources constitute a huge reserve of valuable minerals. For instance, platinum, polymetals, and a number of rare minerals are found in the large quantities of ore in Kolskiy peninsula, Karelia, Norilsk, and Altai. The institute not only possesses exploration, mapping, and deposit estimation capabilities for these sites, it also employs the newest methods for enrichment and physical-chemical processing.

Geophysics

New nonlinear geoelectromechanical methods of exploration are used in the ore and oil and gas fields. For exploration, delineation, and estimation of the ore bodies, the contact method of the polarization curves is used. For the estimation of content of metal in solutions in underground ore leaching, particularly in the copper and uranium fields, the polarographic logging is employed. The concentration of dissolved organic substances, including mineral oil, in underground water and in surface water bodies can be determined by the laser-luminescent logging (LLL). This method is also used to assess ecological conditions, especially along the paths of oil and gas pipelines. An optical system based on the difference between the beam reflections of the surface of clean water and oil film efficiently detects oil leakages from pipelines into rivers and other bodies of water.

Well Drilling

Several types of electrothermal and electromechanical tools have been developed. These tools have been used to bore holes more than 16,000 meters deep in the Arctic and Antarctic to obtain ice cores in the rock iceboxes. For the first time, the cycling of climate change on Earth was proven, and four glacial and interglacial periods were identified as a result of isotope investigations of nearly 600,000-year-old ice cores. Through the use of sterilized sampling in geochemical and microbiological investigations, microorganism in some ice cores were proved to be 200,000 years old. In addition, new techniques using high temperature penetrators simultaneously with ecologically pure pipeless enforcement have been developed for boring holes in loose, loosely bonded, and porous rocks, especially watered and frozen rocks. The newest heat-resistant composites, which do not require the usage of an inert gas to prevent oxidation, also have been adapted for drilling. Electrothermodrilling is important for many projects, including developing water supplies, reinforcing basements of old buildings, and laying cable. A high-temperature penetrator of the condensation type that has been patented in Russia and in the United States could be used for hazardous waste disposal in salt domes. Finally, the institute and Los Alamos National Laboratory are jointly researching methods to enforce and insulate oil and gas wells using bonding materials with low melting points and thermal packing penetrators.

PRODUCT DEVELOPMENT AND SERVICES

In addition to applied research, the institute develops specific products for industry. For the oil and gas industry, the institute is engaged in several enhanced recovery projects, including research on water and steam injections, improved flow rates, and enhanced imaging. The institute also has developed new technologies that are used for surveys of rock massifs, engineering structures, and architectural monuments. These technologies are more accurate and provide more usable algorithms than previous technologies. The effectiveness of the institute's magnetic prospecting and electrometry procedures for studying and extracting underground archeological objects has been confirmed by the discovery of valuable architectural objects in Greece, Italy, and Southeast Asia. Finally, in the area of mining engineering, the institute has developed new explosive materials (EM) for breaking stone blocks and blasting during construction. EM, which is made of available nonexplosive materials, is safe during manufacture, transport, and use; it is also inexpensive. Two other mining engineering innovations are a new air lining for mining flat ore deposits and a mechanical air lining for mining thin beds.

The institute also provides services for industry. For example, the institute's concentrators are used for washing assays, efficiently separating small amounts of heavy metals and minerals from mine refuse, and conducting a variety of tests. In the field of industrial ecology, the institute has developed a system to audit conditions along oil-and-gas pipelines and the state of the environment along the pipelines' path. The institute makes recommendations on construction of ecologically safe underground repositories for the burial of toxic and radioactive wastes. It also has developed several water purification methods.

CONCLUSION

The main research and development activities of the St. Petersburg Mining Institute have been conducted in close cooperation with other universities and with research and production firms, including converted military firms. Most developments are tested and then patented. More than 60 of the institute's patents are in use.

However, as a result of the many years of operating under a planned economy, the institute lacks experience in using its intellectual property in economic activity. Nevertheless its research activities during the past two years have created conditions favorable for the commercialization of such property:

- The ratio of budget to off-budget financing is 1:8.
- A modern computer network has been established.
- A system of social protection for scientists and students has been created.
- New laboratories with modern laboratory equipment have been established.
- Eight scientific laboratories of foreign firms are now located at the institute.
- A special fund for promoting fundamental research has been created.

Under the conditions of economic crisis, science can survive only in those organizations that consider it an economic resource. Science responsive to market demand can create conditions for enhancing the life of scientists and improving their material welfare.

DEVELOPMENT OF LEGAL REGULATIONS FOR TECHNOLOGY COMMERCIALIZATION IN RUSSIA

Vladimir Meshcheryakov
Russian Agency for Patents and Trademarks

Legal regulation of technology commercialization was relatively simple in the former USSR. Economic relations in that period were based on state property and highly centralized production planning. The results of technology workers' intellectual activity were legally protected through certificates of authorship for a given invention, and the certificates gave the state exclusive right to the results.

With the transition to new forms of property in the 1990s, Russian legislation has shifted to the use of patents to protect inventions and other types of industrial property. Under the new system, exclusive rights to the results of intellectual activity are granted to specific individuals or legal entities. The right to obtain a patent on an invention created by a researcher outside the course of his or her employment duties belongs to the inventor or his or her heir. The right to obtain a patent on an invention made in the course of employment belongs to the employer.

All of the special laws on intellectual property in Russia were adopted in 1992 and 1993. These laws include the Patent Law of the Russian Federation, the Law of the Russian Federation "On Trademarks, Service Marks, and Names of Places of Origin of Goods," the Law of the Russian Federation "On Achievements in Breeding and Selection," the Law of the Russian Federation "On Legal Protection for Computer Software and Data Bases," the Law of the Russian Federation "On Legal Protection for Topologies of Integrated Microsystems," and the Law of the Russian Federation "On Rights of Authorship and Associated Rights." The Civil Code of the Russian Federation, which subsequently was adopted, contains general provisions pertaining to intellectual property.

These laws are in accordance with the legislation of highly developed countries in the area of intellectual property. As Russia takes the necessary actions to join the World Trade Organization, it has become clear that no fundamental changes in Russian legislation, especially in the area of industrial property, are required to bring it into compliance with the provisions of the Agreement on Trade-Related Aspects of Intellectual Property Rights (TRIPS). Such compliance is one of the prerequisites for joining the World Trade Organization.

The reliability of the present Russian patent system, and consequently its attractiveness to domestic and foreign investors in the Russian economy, can be evaluated on the basis of two factors: the reliability of the protective documents issued by the patent agency (such as patents on inventions, useful models, industrial samples, and certificates of trademark) and the reliability of the judicial system for stopping violations of exclusive rights in reviewing rights infringement disputes.

RELIABILITY OF PROTECTIVE DOCUMENTS AND THE JUDICIAL SYSTEM

The reliability of protective documents, particularly patents on inventions, was rather high in the former USSR and remains so in Russia today because patent legislation and related administrative directives in the former USSR and now in Russia stipulate very detailed methodological approaches in evaluating the patentability of an invention. Experts from the USSR and Russian patent agency traditionally have performed thorough scientific-technical and methodological analyses of proposed inventions during the course of their patent examinations. This tradition characterized the professional "school" of the Russian patent review process. As a result of changes in patent legislation (for example, the removal of the requirement that a proposed invention have a "positive effect" and the significant expansion of procedural-legal provisions), the experts now devote less time to analyzing the scientific-technical aspects of the inventions. They now focus more attention on analyzing methodological and procedural-legal questions in conducting patent examinations.

In 1996, the Russian Patent Agency (the Chamber of Appeals of the Russian Agency for Patents and Trademarks) reviewed objections to the issuance of 25 patents on various inventions. Of these disputed patents, only three were annulled as mistakenly issued. In 1997 after reviewing objections to 27 patents, again only three were annulled. These decisions on the part of the patent agency were not further challenged in the courts.

The reliability of trademark certificates is another matter. In 1996 after review of objections to 68 trademark registrations, 11 registrations were annulled. In 1997 the complaints against 78 trademark registrations resulted in the annulment of 33 registrations. Only one of all of these decisions on the part of the patent agency was overturned by the courts.

Clearly, the reliability of the judicial system in stopping violations of exclusive rights cannot be deemed as great as the reliability of the patents issued. In the former USSR, the courts almost never reviewed cases involving violations of exclusive rights to inventions or other types of industrial property because exclusive rights to nearly all inventions created by Soviet inventors belonged to the state. Following the passage of the Patent Law of the Russian Federation, which stipulates that exclusive rights to inventions are vested in specific individuals and legal entities, disputes over violations of exclusive rights began to appear. But in connection with the crisis situation in the economy and the low level of production, disputes concerning the violation of exclusive rights to inventions rarely occur. At the present time, judges are only now acquiring the necessary experience to review disputes on violations of exclusive rights, cases which often require specialized scientific-technical and legal expertise. The disputes that currently arise mainly involve violations of exclusive rights to trademarks and especially copyrights (as with works of science, literature, and art). The special arbitration courts that have been created in Russia are a significant help in providing professional, high-level review of court cases involving violation of exclusive rights. In contrast to general jurisdictional courts, these courts only review disputes of an economic nature, including disputes concerning violations of exclusive rights.

The Russian patent system provides rather firm guarantees of observance of the rights of investors to intellectual property they use, particularly in the case of inventions, as long as the investors themselves make professionally competent efforts to obtain legal protection for this intellectual property in Russia. Specifically, they must ensure that applications for patents on inventions are prepared in a professionally competent manner in accordance with Russian patent legislation.

Russia, like many countries, has problems with the violation of exclusive rights to objects under copyright and associated rights, including works of science. Copyrights are appropriate in cases in which a work does not require a patent or other expert examination or state registration. In contrast with inventions, these works are more difficult to identify; therefore, violations of exclusive rights to them are more difficult to stop. For example, the so-called "audio and video piracy" problem exists throughout the world. Pirated audio or video cassettes are rather easy to produce and distribute, and their noticeably lower cost compared to licensed audio and video products is very attractive to people whose are not well-off financially.

Russia currently is making legislative changes to curb copyright violations. Judicial legislation must stipulate accelerated procedures for court review of lawsuits concerning violations of exclusive copyright and associated rights. Customs legislation must set forth the right of customs agencies to seize counterfeit products passing through customs borders when they receive a complaint from the copyright holder, as stipulated by the provisions of TRIPS.

RIGHTS OF THE STATE TO THE RESULTS OF FEDERALLY-FUNDED WORK

Despite the merits of Russian intellectual property legislation, special Russian laws in this area, particularly the area of inventions, have failed to address the rights of the state to the results of intellectual activities performed with state funding. In particular, the Patent Law does not provide for the possibility of granting the state or its representative a patent for an invention created using funds from the federal budget. Moreover, this law does not stipulate any rights to such inventions on the part of the state. Any commercialization of technologies on behalf of the state or its interests is complicated when these technologies are based on inventions created using federal budget funds because the patents for the inventions are issued to specific individuals or legal entities.

In Russia highly science-intensive technologies are created with federal budget funds under state programs and on the basis of state contracts. The contracts are for scientific research, experimental design, engineering work, and the provision of goods to meet federal government needs. Government customers in these contracts—federal executive agencies, federal fiscal enterprises, and state institutions—are representatives of the state itself. Problems concerning the state's rights to the results of intellectual activity conducted during work on state contracts can be resolved by including in the contracts the appropriate provisions granting to the state in the form of its representatives the necessary authority to commercialize any technologies developed. However, effective resolution of these problems on the basis of state contracts is hindered by insufficient legal expertise and experience in concluding such contracts.

The aforementioned condition of the legislative base has led to the spontaneous redistribution of rights to the results of state-funded intellectual activity even in the Soviet period. In the process, controversial and undefined legal relations have arisen in connection with the failure in many cases to observe procedures for formalizing rights to such results. All of this hinders the normal process of commercializing these results.

To further stimulate investment in the Russian economy and promote the growth of industrial output, measures to resolve the aforementioned problems in the area of technology commercialization are planned. The Ministry of Science and Technology of the Russian Federation and the Russian Agency on Patents and Trademarks, in cooperation with other federal executive agencies, have prepared drafts of normative legal acts calling for the establishment of state policies to develop the intellectual property market and bring the results of scientific-technical activity to bear in the national economy. These documents set forth the main points of the policy, which would include provisions for:

- a balance of legal interests among subjects of legal relations, including the state, in the creation, legal protection, and use of objects of intellectual property and other results of scientific-technical activity;
- state support of processes involving the creation, legal protection, and use of the results of scientific-technical activity and an increase in the competitiveness of products made by domestic manufacturers; and
- material support for the authors of domestic scientific and technical innovations that adequately reflects their contribution to the socioeconomic development of Russia.

Under this policy, the procedures for using the results of scientific-technical activity obtained during fulfillment of state contracts would stipulate the distribution of rights to these results to state customers as representatives of the state and to the contract performers (product developers and manufacturers). Because the issue of acquiring exclusive rights to an invention arises during the performance of scientific research and experimental design work, the terms of a state contract must include provisions outlining the right of the developer to obtain a patent on inventions that he or she creates while working on the contract.

The contract terms must also include limitations on the rights of the patent holder to ensure that the exercise of these rights is coordinated with the actions of the state customer. For instance, if the state places an order for the manufacture of products involving the use of an invention to which the contractor holds a patent, the contractor must at the request of the state customer grant the manufacturer of the products a non-exclusive royalty-free license. This precludes the possibility that the state would pay first for the development of a scientific-technical product and then for the right to use the results in the interest of the state.

The state also must have the right to consider the interests of the state and society when commercializable technology has been developed within the framework of a state contract. For example, if necessary, a product produced under a state contract might first be sold on the domestic market to meet the needs of the state and society and then be sold abroad. Furthermore, the income received as a result of the sale of the products must be fairly divided between the creator of the scientific-technical results, the manufacturer of the product using these results, and the state. Finally, state contracts must spell out budget expenditures for the patenting of inventions in Russia and in foreign countries where there are markets for products based on these inventions.

In addition to measures connected with state contracting procedures, efforts are under way to amend the Patent Law and other special laws providing legislative protection of the rights of the state to the scientific-technical results of federally funded work. The Russian Agency for Patents and Trademarks already has prepared a draft law on changes and additions to the Patent Law. The draft law has been submitted to the government for subsequent coordination with interested federal executive agencies and review by the State Duma.

A number of financial issues also must be addressed. For example, appropriate changes in the tax law also must made to reduce the tax burden on enterprises that manufacture products using patented inventions and other types of intellectual property. Also needed are better rules for cost valuation of intellectual property. This valuation is necessary for the accounting of an enterprise's nonmaterial assets. The value of intellectual property is particularly important in the privatization of enterprises, as well as in the handling of court cases involving the violation of exclusive rights of a patent or copyright holder. Development of procedures for cost valuations of intellectual property, a process for training and licensing individuals to make such valuations, and a system for including intellectual property as part of enterprises' non-material assets is planned.

RIGHTS OF PARTIES TO RESULTS OF JOINT EFFORTS

Recently increased attention has been paid to the regulation of the rights of each party to scientific-technical results from projects involving foreign partners. The Government of the Russian Federation has concluded a fairly large number of agreements with the governments of other countries regarding scientific and economic cooperation. Requisite conditions of these agreements are provisions regulating the rights of parties to the scientific-technical results of collaborations. These provisions reflect consideration of mutual interests and the national legislation of the parties, including international treaties the parties have signed.

The provisions may include recommendations to specific organizations participating in joint enterprises (henceforth to be called "Participants"). Participants in agreements on joint work define all intellectual property for which the creation, use, and transfer is reasonably foreseen during execution of the agreements. Such intellectual property is categorized either as previously existing or as newly created. Agreements should indicate that previously existing intellectual property may be used only after actions are taken to ensure the necessary legal safeguards.

In the section of agreements dealing with the distribution of rights to jointly created intellectual property, the participants must take into account various factors, including the contribution of each participant to the work performed (such as previously existing intellectual property, intentions, commitments, and capability to provide the necessary legal protection for jointly created intellectual property) and the proposed participation of each participant in commercial use of jointly created intellectual property. Moreover, if necessary agreements should indicate the expected recipients of such property, the types and extent of use of the property in the territory of each party and in other countries, the extent of use of previously existing intellectual property, the rights of the participants to act on confidential information and their responsibilities to protect it, and the rights of each participant in the event that the other participant does not meet its obligations to ensure protection for the intellectual property. Agreements also should include provisions and procedures for payment of compensation to inventors and authors of other types of intellectual property.

Agreements should cover procedures for submission of patent applications to the national patent agencies of each party involved. Patent applications concerning inventions created on the territory of a given party should first be submitted to the patent agency of that party's country.

Agreements also should address that the transfer of assets involving elements of intellectual property from one party's country to another party's country for the purpose of joint activities under the contract. Furthermore, agreements should specify that this transfer must not violate the legal rights of any third parties in the country from which the transfer is made. They also should specify that any complaints lodged by third parties regarding the transfers will be the responsibility of the participant which made the transfer.

APPLICABILITY OF THE U.S. EXPERIENCE WITH TECHNOLOGY COMMERCIALIZATION

The U.S. technology commercialization experience with which the Russian delegation familiarized itself appears in many respects to be applicable to Russia. For instance, as a result of the Bayh-Dole Act, U.S. patent legislation allows the rights to inventions created by national laboratories and universities to be acquired not only by the U.S. government, national laboratories, and universities, but also by nongovernmental investors who have provided funds for the creation and commercialization of technologies. As noted above, the Patent Law of the Russian Federation, which stipulates that rights to inventions created using federal budget funds be assigned to any individuals invested in the creation and commercialization of the technology, does not include the possibility of assigning rights to these inventions to the state. For this reason, the Patent Law of the Russian Federation is not as universal as the patent legislation of the United States.

The United States has experience in creating the necessary conditions for financing the entire process of producing and utilizing technologies, including conducting basic research and experimental design work, preparing test models (prototypes), launching mass production, and selling the products. Unfortunately, application of this experience to Russia largely is hindered by the economic crisis, which has resulted in limited willingness to use new technologies, and by the insufficiency of state financing.

COMMERCIALIZATION FOR THE POLYMER INDUSTRY: THE EXPERIENCE OF AN ACADEMY INSTITUTE

A.N. Ozerin
Institute of Synthetic Polymer Materials

AREAS OF SCIENTIFIC ACTIVITY AT THE INSTITUTE

The Institute of Synthetic Polymer Materials (ISPM) is part of the Russian Academy of Sciences Division of General and Technical Chemistry. The activities of the ISPM conform to the conditions set forth in Point 3 of the Statute on State Accreditation of Scientific Organizations: scientific and scientific-technical activities are fundamental at the institute; the volume of such work totaled 100 percent of all work performed at the institute during the past three years; and the institute's charter provides for a scientific council to serve as one of its administrative organs.

Research at the ISPM mainly focuses on the creation of fundamentally novel polymer materials and composites and the development of technologies for producing them. In recent years, the institute has:

- developed models for a process of high-temperature multiple fragmentation of multi-component polymer systems—models important in the creation of new environmentally safe technologies for producing composite polymer materials;
- discovered a "macromolecule-particle" and devised fundamentally new methods for synthesizing dendrimer and superbranched organosilicate macromolecules—a new class of functional polymer structures with broad applications;
- developed principles for creating heavily loaded systems that efficiently absorb electromagnetic radiation in the super high-frequency spectrum; and
- developed a technology for producing a wide range of composite materials using secondary polymer materials, including composites with flame-resistant properties.

With its unique processing and research equipment, which meet world standards in chemical engineering, the institute can conduct a full array of research on the structure and properties of polymer materials and composites.

With its highly-qualified scientific personnel (110 researchers, including one member and one corresponding member of the Russian Academy, 10 doctors of science, and 51 with the candidate of sciences degree [equivalent to the Ph.D.]), along with the theoretical, computational, and experimental methods at its disposal, the institute is capable of solving basic and applied problems concerning the physics and chemistry of polymer materials. In addition, the institute educates scientific personnel through its professional training system. As part of this system, the institute has established a special on-site Polymer Physics Department at the Moscow Physical-Technical Institute (MPTI). Approximately 20 students are educated and receive specialized training at the department each year. The department's graduate school offers degrees in three fields. Among the institute's staff are 18 young scientists (under age 33) and 15 graduate students.

The institute maintains strong scientific ties with institutes of the Russian Academy of Sciences (RAS); universities; industrial or sectoral institutes and design bureaus; and foreign scientific organizations such as the University of Ulm (Germany), Chalmers University (Sweden), and the Dow Chemical and Armstrong companies (United States).

For a long time, the activities of the institute were directed toward research to create fundamentally new types of polymer materials as part of the overall state plan for scientific and technological development. This work was performed in accordance with the RAS research plan, which had been worked out by the country's planning agencies. Given this organization of scientific activity, commercialization of research was not a task for the research institutes themselves. Instead, the end results of scientific research were passed on to the industrial scientific production centers, which had their own experimental production bases to conduct testing and design work and to develop concepts to the level of industrial technology demonstration projects, at which point the finished technologies were handed off to industry.

With the process from scientific development to industrial production organized in this manner, the question of property rights to newly created scientific products did not arise as the results of the work of all those involved in the process belonged to the state. Moral incentive was the main factor giving researchers an interest in creating new scientific products. However, the scientific collectives that proposed new developments, and thus displayed their high creative potential, received additional support from industrial centers. This support took the form of contracts for research on the technological issues that these centers addressed.

With the transition to a market economy, the scientific organizations of the Russian Academy of Sciences face two major challenges. The first is identifying new areas of strategic development. The second is attracting sources of financing to supplement funds from the state budget, thereby ensuring that research meets world scientific standards.

Under current economic conditions, a very important element has disappeared from the process leading from scientific development to industrial production—the industrial research and production centers. As a result, society's demand for science in Russia has diminished substantially. Now scientists rarely gain the satisfaction of seeing their discoveries employed for the public good, further reducing their incentives to create new scientific products. To maintain their scientific potential and find support for their scientific activities, research groups at RAS institutes have three options: (1) participate in various competitions, including international competitions, to conduct research; (2) solicit the involvement of foreign industrial firms in cooperative activities; and (3) attempt to sell the results of their scientific activity.

Financing under the first option is directed entirely toward support of basic scientific research. Such support is unlikely under the second option because foreign industrial firms are not inclined to make significant material investments in basic research in Russia. Typical research contracts have an average value of $10,000–$15,000, and a maximum value of $60,000. As a rule, the firms insist in the contract that they maintain exclusive ownership rights to any intellectual property created in the process of work on the contract. Nevertheless, the funds received by scientific groups working under contract to foreign firms represent an important additional source of financing, allowing these groups to satisfy their day-to-day needs, such as purchasing reagents from abroad or repairing imported scientific equipment. Therefore, such contracts are rather attractive to RAS institutes.

Financing under the third option, commercialization of scientific developments, is potentially the most significant for RAS institutes. To a large extent, such financing can and should be directed toward the support of basic scientific research, thereby creating material incentives for industrially oriented development work. However, this option is the most difficult to implement. This paper will examine typical problems that the RAS Institute of Synthetic Polymer Materials (ISPM) continues to face in the practice of commercializing its research products.

PRACTICE OF PLANNING SCIENTIFIC RESEARCH

Senior researchers in Russia's scientific institutions (including the ISPM) possess considerable experience in practical scientific activity. The heads of laboratories or departments often have 30 or more years of seniority. As a rule, scientists conduct many years of research in the same scientific institution where they began work immediately after graduating from a university or institute. Long years of experience in practical work in a given scientific field enable institute section leaders to follow trends in the development of world scientific thought rather closely and to plan their own promising research accordingly. Therefore, the lab director or department head is an expert on any new scientific task at hand and often one of the most qualified researchers to deal with it. To ensure a high level of leadership for scientific research, these manager are periodically certified and must be re-elected (usually terms are five years). In addition, the significance of the work performed by their research groups is evaluated by the broad scientific community.

While many excellent senior scientists remain active, ensuring a constant stream of young specialists into the research sphere is a problem. It has become a particularly urgent concern because the attractiveness of scientific work among young people has declined noticeably in recent years. Thanks to its on-site department at MPTI, the ISPM successfully attracts undergraduate and graduate students to assist in its scientific work. These students have a good basic education in physics and mathematics and a mastery of current computer technology.

Young specialists trained at MPTI face a lengthy period for preparation of the master's thesis. Students come to the base institute (one of which is ISPM) after three years of preparation at MPTI in the basic disciplines: general and applied chemistry, physics, mathematics, and two foreign languages of their choice (English, German, French, or Japanese). During their first year of study, the students carry out projects at the base institute (ISPM) and receive bachelor's degrees. During the next two years, they continue working on their areas of specialization, after which they receive master's degrees. Having provided three years of instruction, the base institute (ISPM) has an opportunity to hire excellently trained young specialists who are well acquainted with the problems and tasks they will face in the future. The young specialists also are given the opportunity to continue study at the graduate level and obtain the candidate of sciences degree. During their next three years of study, students spend an increasing amount of time at the ISPM. In this way, the student's research supervisor—who as a rule is a professor at the on-site department at MPTI and simultaneously head of a research lab at ISPM—has the opportunity to work constantly with the young specialist during the three (or six) years of his or her course of study. MPTI undergraduate and graduate students are critical to the development and adaptation of modern research methods and facilitation of cutting-edge research at ISPM in the newest scientific areas. Their work is very highly regarded throughout the Academy of Sciences, and ISPM attaches great significance to this important source for qualified young specialists.

One important element in selecting long-term research objectives is information. Scientific groups therefore regularly track information published in the open press. In the field of polymer materials, the most informative and popular publications among Russian researchers are *Chemical Abstracts*, *Macromolecules*, *Trends in Polymer Science*, *Polymer Symposia*, *Modern Plastics International*, and *Polymer News*. Information accessible on the Internet has taken on special significance in recent years. Unfortunately, long-term planning of scientific research is hindered by the limited number of scientific periodicals received by the country's main scientific libraries in recent years and the unsatisfactory level of development of the Internet in Russia. One way to solve this problem might be state support for major scientific-technical libraries and the accelerated development of the Internet for state-financed organizations, including RAS institutes.

For long-range research planning, periodic analysis of global trends in natural science disciplines (physics, chemistry, biology) also would be exceptionally useful. The task of compiling such forecasts might be assigned to the scientific councils for the various specialties represented in the RAS divisions. The coordination and support of the Russian Ministry of Science and Technology would facilitate this work.

The existing system for evaluating world markets in order to plan scientific products also should be considered completely unsatisfactory at present. A possible solution to this problem would be the creation of a network of state (or international) centers to study market conjuncture and demand in the most important sectors of industrial production.

The efforts noted above would significantly enhance the quality of research planning at RAS institutes and would facilitate the efficient use of federal budget funds.

The scientific developments considered below were the result of a comprehensive program of basic scientific research, planned in the late 1980s, on "Environmentally Safe Chemical Processes and Technologies." Research areas were chosen by the ISPM scientific collective under the leadership of Academician N.S. Yenikolopov. In this case, the deciding factor in the selection of these areas was the practical experience and intuition of the scientific group and its leader, not the importance of the problem, which had yet to be established at the state level. By 1994, a demonstration of the technological potential of the institute's research results and the commercialization of these results was possible.

CONDUCTING SCIENTIFIC RESEARCH

The results of ISPM's research on the properties of solid-phase chemical processes in polymers under mechanical stress could be successfully put into practice on an industrial scale. These results include the discovery of highly efficient and environmentally friendly methods for the high-temperature fragmentation of plastics and rubbers, the development of principles for using these materials in the creation of various composites, and the development of prototypes of equipment for carrying out these processes.

In pursuing this particular program of experimental research, ISPM encountered no great difficulties and was given sufficient financing because the research was included in the ISPM's work plan. At the next stage of the work, additional funds were obtained on a competitive basis from the Federal Targeted Scientific-Technical Program "Environmentally Safe Processes for Reprocessing Secondary Polymers." These funds played a decisive role in moving the project forward. In the final stage of the project, the level of this support equaled that provided by the state.

On its own, ISPM could not identify all of the potential areas of application of the processes and materials it developed. One reason is that the institute lacks the equipment necessary to conduct comprehensive research and testing of new materials. One possible solution to this problem might be for the state to support the network of collective-use centers created recently by the Russian Foundation for Basic Research to facilitate work on grants for basic experimental research. Such centers make it possible to use unique and expensive research equipment under favorable terms to work on scientific projects financed by the state budget.

EVALUATING THE SIGNIFICANCE OF RESULTS OBTAINED

The ISPM encountered serious difficulties in evaluating the practical significance of the scientific results noted above. In general, the economic assessments were made by the members of the scientific group. The information used for these assessments was obtained by analyzing materials published in scientific periodicals, evaluating the current patent situation in a given field, and talking with colleagues, including foreigners visiting the ISPM. Such an analysis is obviously not comprehensive, and the conclusions can be implemented only to a limited extent.

Such analyses should be performed by specialized state centers designed to study potential demand for these products, because the customer and end-user of work carried out under various environmental programs is the state. The need to create such centers was noted above. The centers could be equipped with small experimental production facilities capable of producing representative sample batches of new materials to be sent to potential customers for testing. The centers might organize permanent exhibitions of new research developments, which could be of substantial assistance in the search for investors for the industrial production stage. The necessary regulations for cooperation between the centers and the institutes would need to be worked out.

COMMERCIALIZING SCIENTIFIC DEVELOPMENTS

Many patents have been granted on new materials and processes as a result of past scientific developments at ISPM (SU Patent Numbers 1653281, 1703468, 1655008, 1669933, Applications Number 96108551/04 [014070] and 96122084/04 [028755]). Through licensing of patented developments, production of a multilayered material for use in protective coatings on buildings and structures has been initiated, and technologies for fragmenting and separating mixed industrial and domestic polymer wastes have been implemented at enterprises in the city of Moscow and in other regions of Russia.

The successful commercialization of scientific developments has become possible thanks to the convergence of interests among the researchers working on the problem, the institute where the work is performed, and the enterprise that wishes to implement the research concept in industrial production. Existing patent legislation makes it possible for the parties to settle issues of intellectual property rights. However, monitoring and enforcement of agreements have not yet been fully established, nor has a mechanism for realizing the rights of intellectual property owners. In essence, agreements are really only "gentlemen's" agreements, and mechanisms for resolving disputed situations remain an open question.

ISPM has encountered serious problems with commercialization of its scientific developments. Given current Russian economic conditions, the active participation of inventors at all stages of the process is needed to bring an idea to the industrial production stage. A reasonable balance between the interests of the individuals involved and the institute as an organization also is needed. The lack of a Law on Inventions Made in the Course of Employment and the absence of needed changes in the Patent Law on State Ownership Rights to Intellectual Property seriously complicate the establishment of such a balance.

One question that has arisen in determining the cost of a license for a process or material or in organizing a joint production venture with foreign partners is the process to be used to establish the value of expenditures incurred by the Russian side. Analysis of this question is needed at the state level.

CONCLUSIONS

The practice of commercializing individual scientific developments at ISPM shows that these developments are sufficiently competitive in the market for scientific-technical products and can be brought to the industrial production stage. Nevertheless, the institute's experience is more along the lines of the experience of mistakes instead of the experience of successes. Under the crisis conditions facing Russian science, successful commercialization of a scientific development is more the exception than the rule. Scientific developments are not "launched" into a market which is ready to adapt them immediately, but rather these developments are slowly "pulled" into the market. The only positive point in this process is the rich experience and knowledge of all characteristics of the modern technology market that institute staff members are acquiring.

Many external factors hinder the successful advancement of ISPM scientific developments from reaching the market. The following actions would accelerate substantially the movement of the institute's scientific developments to the commercialization stage:

- effective state support for major scientific-technical libraries and accelerated development of the Internet for state-financed organizations, including RAS institutes, to ensure that these organizations have full access to information about scientific developments during the research planning stage;
- periodic publication of forecasts of global trends in the natural sciences (physics, chemistry, and biology) to facilitate long-range planning of scientific research;
- creation of a network of state (or international) centers in Russia to study market demand in the most important sectors of industrial production;
- state support for the network of collective-use centers recently created by the Russian Foundation for Basic Research to facilitate work on grants for basic experimental research; and
- immediate adoption of the Law on Inventions Made in the Course of Employment and changes in the Patent Law on State Ownership Rights to Intellectual Property.

THE MAIN PROBLEMS IN COMMERCIALIZATION OF SCIENTIFIC RESEARCH RESULTS

A.P. Simonov
Karpov Institute of Physical Chemistry

INTRODUCTION

Commercialization of the results of scientific research is an old problem. However, the transformation of a scientific idea into a product for industrial use (see Fig. 1) very often is interrupted during the first three stages of commercialization. Engineering is the most difficult stage of this process. At this stage, scientific results must be transformed by engineers into a real industrial process that can be shown to be efficient and profitable enough for companies to buy. This stage is quite expensive, especially in the chemical field. External funding often must be found to cover because, as a rule, research institutes do not have sufficiently large capital resources.

In the USSR, research and engineering costs were covered by the state. However, because of shortages of money, even the best research results often could not be properly introduced into industry. Thus, most research ended only in patent applications, publications, and reports to the organizations that had ordered the work. Chemical plants supported some research projects, but they usually had strict plans for production and for products and were not interested in innovations.

In Russia, scientists in state research institutions now encounter the same problem of capital shortages and insufficient government support. Privatized institutes have sharply reduced their investigations or changed the direction of their activities. The chemical industry remains stagnant, and chemical companies are only beginning to show some interest in improvement of the technologies they use. Therefore, investors or sponsors must be found to help research institutes complete the engineering stage of product development.

In principle, the results of applied research or even an idea can be sold at any stage of their development, although selling results when they are ready to be used in industry is much more profitable. An institute that is able to sell the final product of its scientific activity, namely goods produced by its own plant using its own technologies, can obtain the highest possible profit. Some attempts

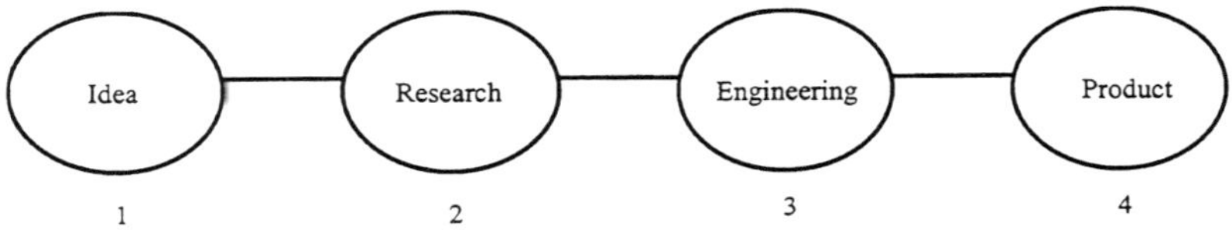

FIGURE 1 Transformation of a Scientific Idea into a Product for Industrial Use

by the Karpov Institute of Physical Chemistry to commercialize research results at different stages of development are described below.

EXPERIENCE OF THE KARPOV INSTITUTE

The Karpov Institute was founded in 1918 as an applied research institute but later became largely an academic organization carrying out both basic and applied research in most areas of modern physical chemistry. It is located at two sites in Moscow and has a branch in nearby Obninsk. The total staff numbers about 1,800, including 800 scientists. The Obninsk branch has been oriented mostly toward applied research and technology development, especially in the field of radiation chemistry. The institute's total budget in 1997 was about $6 million, of which the state provided $2 million and grants and contracts provided $4 million. Most of these funds were earned from the sale of products produced in Obninsk with the institute's own technology and facilities. Table 1 lists these products. Only products produced under a specific license may be legally sold.

The institute's research nuclear reactor is in Obninsk. Fifteen years ago, in an attempt to use this expensive apparatus more efficiently, the decision was made to develop chemo-nuclear technology for obtaining radioactive medicines for diagnosis and therapy and to sell medicines rather than technology. With capital provided by the state, several new technologies have been developed, leading to industrial production of a wide variety of bioactive substances containing molecules marked by the radioactive isotopes of Tc-99m, 1-123, I13 1, Sm- 1 53, Re- 1 86, and W- 1 88. These radiopharmaceuticals are sold to more than 290 state and private hospitals and clinics all over Russia. A considerable portion of the profits is spent on scientific investigations of the radiopharmaceuticals, a search for new medically selective molecular carriers of radioactive isotopes, development of chemical synthesis and purification methods, medical testing of new preparations, licensing of industrial production

TABLE 1 Radiopharmaceuticals Developed at the Karpov Institute

Isotope	**Chemical Form**	**Production Start**	**Application**
Mo-99	Tc-99m generator	1988	Diagnostic
I-131	Sodium iodide in isotonic solution	1990	Thyroid gland diagnostics and therapy
	Sodium iodide in capsules	1998	Thyroid gland diagnostics and therapy
E_{γ}=0.365 MeV	*Ortho*-Iodohippurate of sodium	1995	Kidney diagnostics
	Rose Bengal	1995	Liver diagnostics
$T_{1/2}$=8.04 day	Albunim	1997	Hemodynamics diagnostics
	Macroaggregates of albunim	1998	Lung diagnostics
	Meta-Iodobenzyl-guanidine	1998	Adrenal gland therapy
I-123	*Ortho*-Iodohippurate of sodium	1998	Kidney diagnostics
E_{γ}=0.159 MeV	*N*-isopropyl-para-Iodamophetamine	1998	Cortex diagnostics
	Meta-Iodobenzyl-guanidine	1998	Miocarditis diagnostics
$T_{1/2}$=13.3h	(para-Iodophenyl)-ß-methyl-pentadecanoic acid	1998	Heart diagnostics
Substances for chemical kits	Oxabiphoric	1993	Skeleton diagnostics
	DMSA	1994	Kidney diagnostics
Tc-99m radio-pharmaceuticals	Bromezida	1995	Liver and gall bladder diagnostics
	Tetraphosmin	1998	Hear diagnostics
Sm-153	Oxabiphor complex	1998	Skeleton therapy
Rc-186	Microspheres of Albumin	1998	Joints therapy
W-188	Re-188 generator	1998	Joints therapy and diagnostics

of medicines, technology improvements, and limited research not directly connected with the field of radiopharmaceuticals.

Other results of investigations in the field of radiation chemistry are used to produce semi-industrial goods: high-quality polymer filters for filtering water, juices, beer, and alcoholic drinks; high-voltage electric insulators; latexes for washable wall papers; and foam polymer materials. However, compared with the radiopharmaceuticals, these goods are produced on a much smaller scale and garner less money.

Perhaps the best way for an institution to commercialize its research results is to produce products with its own large-scale production plant using its own patents, technologies, expertise, and other resources. But this approach has its drawbacks. First, not every institute has its own large-scale production plant. Second, large-scale plant operations can divert resources, including human resources, from research activities, especially as specialists engaged in business-oriented endeavors usually receive much higher salaries than scientists engaged in basic research. Finally, research institute personnel may not have the appropriate skills to determine market demand for products.

Another way for an institute to commercialize its research results is to act as an engineering firm for a company. Using existing technology that it had improved, the Karpov Institute reconstructed the ammonia production line at one of the ACRON chemical company's plants to produce methanol. All expenses, including payment for the license for the initial technology, were covered by ACRON. But further improvement of the production technology was possible only through the use of institute's research and development. ACRON suggested that the institute pay to start a pilot plant and test the process. In the event of success, ACRON would repay the institute's investment and give the institute a portion of the plant's profit from the technology during a specified period of time. The institute agreed to this arrangement. In this particular case, the institute's expenses were not large and no money needed to be borrowed. If special funds existed to support engineering firms in the demonstration and testing of new or improved processes, similar arrangements could be widely used.

In another attempt to commercialize its research results, the Karpov Institute constructed a fairly compact and automatically operating apparatus for production of hydrogen from natural gas, an apparatus convenient for users of both natural gas and hydrogen—such as electric power stations. To date, the institute has sold only one such device, which has been operating successfully at a Moscow power station for two years. The sale of additional units has been hampered by a shortage of capital for proper advertising of the product.

Obtaining large bank credits to compensate for a deficit of capital is difficult and dangerous for an institute. It is difficult because, in general, banks pay little attention to scientific research and commercialization of its results, however sound the business plans. During the ongoing privatization process in Russia, banks are keen to acquire property. Obtaining large bank credits is dangerous because loans must be repaid in a short time, typically one year. In addition, bank interest rates are very high (now 36-40 percent). The ability of institutes to repay money spent on the development of a chemical process or construction of equipment depends on the financial state and stability of the companies on whose behalf such investments have been made. In this respect, foreign customers are in most cases more predictable and reliable than domestic ones and therefore are more attractive to research institutes. Moreover, foreign customers are easier to find and to strike a bargain with than domestic customers. Finally, obtaining bank credits is difficult for institutes because the terms of a deal between an institute and a bank are always the result of negotiations. At the present time, there should be preferential treatment for bank credits for research and development, including lower interest rates, longer-term credits, and state guarantees for debt repayment.

Another approach to commercialization is the sale of research results that require additional investigation and testing to an independent engineering firm. The sale of semi-products allows buyers to offer much lower prices, to require exclusive rights on further refinements, and to stipulate strict conditions for future work. The seller has no choice but to accept almost every demand.

Even when an institute successfully develops a commercializable product, further problems may arise. For example, the Karpov Institute's scientists developed a new electrode material for electrochemical synthesis of sodium chlorate and other oxidants. The electrodes, which contain much lower quantities of precious metals and last much longer in corrosive media than commonly used electrodes, were insufficiently tested in real industrial conditions. A Canadian chemical engineering company wanted to buy a license to produce and sell these new electrodes. However, the invention had been patented only in Russia, a situation that some western countries consider to be equivalent to the premature publication of an invention. Insufficient capital prevented the Institute from obtaining a patent at the right time and in the necessary number of foreign countries. As a result, the Canadian firm experienced difficulties when, in accordance with the licensing agreement, it attempted to patent the electrodes in the United States and other countries. These additional expenses for patenting and testing resulted in lower licensing payments to the institute. The institute's experience suggests that the state should support patent activity by creating a special fund that would provide low-cost loans to cover patent fees.

Attempts to commercialize initial research results also can be problematic. The Karpov Institute's contracts with some large foreign companies, such as FMC, Hughes Aircraft, Bayer, and Haldor Topsoe, call for no distribution of intellectual property rights or of profits after commercialization; the firms retain these rights and profits. Further, no reward for investors is stipulated. The firms have strict conditions for the research and require permission for publication of research results. Although the institute attempted to prepare the texts of contracts in accordance with the recommendations issued on July 16, 1996 by the Joint Russian-American Commission on Economic and Technological Cooperation, it eventually signed documents that appear to be unfair. To illustrate, here is a citation from a document the Institute signed with an American firm:

> "Karpov Institute of Physical Chemistry (KIPC) agrees that any data and technical information (a) obtained directly or indirectly from the firm in connection with this agreement or (b) related to or derived from work on projects performed by KIPC under the terms of this agreement is the confidential property of the firm, and the firm has and shall have the right to exclude its use by others including KIPC......"

Further,

> "KIPC agrees that it will be performing project work as requested by the firm and agrees to assign to the firm all rights, title, and interest in and to any idea, invention, know-how, trade secret, and improvement, which is conceived, discovered, or developed as a result of the work performance under this agreement. This term is to be effective for a minimum of 12 years from the completion date of the project work."

The terms of contracts with domestic customers are less strict. However, these terms do not mean that domestic partners are more friendly. Rather, they reflect these partners' more limited market experience.

Although the conditions of any deal are the result of discussion and consensus, some legally determined limits should apply to the basic terms of documents on cooperation, licensing, and confidentiality agreements. For example, the time period for exclusive rights of one of the partners should not be longer than three to five years. An award for inventors should be obligatory. The inventors as well as the institute where they work should receive a portion of the profit gained after the completion of the work. Former Vice-Premier V.B. Bulgak suggested that Russia form an organization to work out, on the basis of international practice and experience, legal standards for contracts involving Russian research institute—standards that would protect the rights of Russian scientists.

Preparing and signing a contract in accordance with the recommendations mentioned above is much easier when the partner is a public organization, such as the National Science Foundation, the U.S. Civilian Research and Development Foundation, or the International Science and Technology Center (ISTC) (which was founded by the American, European, Japanese, and Russian governments to offer civil research opportunities to scientists who earlier had been engaged in military investigations). According to ISTC rules, the participants in a project financed by the ISTC are the owners of the intellectual property created during work on the project. These participants can submit patent applications when and where they wish. The inventors obtain a portion of any payments for implementation of the project. A Karpov Institute project that is financed by ISTC (#193: Design and study of new radiation stable materials for use as scintillators for radiation control) recently has been finished, and the

ISTC will pay for three patent applications. Moreover, the ISTC will help find potential users of the patents.

The experiences described above indicate that the main problem in commercialization of valuable scientific results is the severe shortage of capital at Russian research institutes and low investment activity in Russia. Other problems include a lack of qualified managers and experienced specialists in marketing technology, an absence of legal protection of technology, and inadequate laws regulating relations in intellectual property and patents. Nevertheless, money problems are the most serious at present.

Perhaps the best research product to market is a license. A license reflects both the novelty of a technology (payment for a patent) and its scientific complexity (payment for know-how). A license agreement must be formulated very carefully. The Karpov Institute failed to sell a good license because it had not included in the agreements some provisions later found to be quite important. Among these provisions were (1) a business plan for the transformation of the invention into a commercializable technology with the buyer being responsible for fulfillment of the plan; (2) a stipulation that the inventors would participate as supervisors and consultants in this process; and (3) a list of countries where patents for the invention would be obtained and an indication of the time period for and costs of obtaining these patents.

Typically the buyers of licenses are engineering firms that introduce technology into industry. The main portion of the license payment is provided by the customers who contract with engineering firms to introduce technology into their enterprises. Therefore, market demand for licenses strongly depends on investment activity. Today in Russia, the number of engineering companies is small and investment activity is low. Therefore, the sale and implementation of good licenses are in a depressed state. Normalization of the situation might be achieved, in part, by the creation of science and technology funds.

SCIENCE AND TECHNOLOGY FUNDS

According to statistical data for 1996, approximately 61 percent of expenditures on research and development in Russia were covered by the federal budget, 27 percent by research institutes and organizations in the business sector, 6 percent by foreign sources, and 6 percent by funds "independent" of the federal budget. These expenditures were mostly for research, not for engineering and implementation of technologies. A few well-known governmental funds, such as the Fund for Technological Development at the Ministry of Science and Technology, support of innovations in all industries. Perhaps several dozen small and practically unknown government funds are very narrowly oriented. Both kinds of funds are almost inaccessible. Equity funds in Russia are very limited.

The funding crisis faced by research institutes that attempt to commercialize their achievements could be mitigated by science and technology funds (STFs) for implementation of technologies based on scientific research results. Domestic and foreign industrial companies and individual entrepreneurs, as well as the state, should establish STFs. Because funding typically is allocated to novel technologies that require no more than $0.5 million for development, an STF with starting capital as low as $2 or $3 million could have a significant impact. The state should stimulate formation of STFs by taxation policy. A portion of the profit of a shareholder of an enterprise would be transferred to the STF before any tax on the profit has been paid. The total sum of this tax would constitute the state's share of the STF's capital.

STFs would provide loans at reduced interest rates to engineering firms that build and operate pilot plants based on new, cost-effective, and patented processes. The investor would repay the loan in full after the pilot plant has reached the expected project yield. Alternatively, the investor would repay the loan when profits are sufficient to cover the investor's expenses.

The state, like any other STF investor, could sell its shares in an STF after a specified time. Before the state's exit from the fund, the profit gained by the STF would be taxed only when distributed but not in capitalized form. STFs should receive priority from the Export-Import Bank of Russia for financing their projects. The credit resources of STFs should be distributed on the basis of competition in which potential investors would participate. To further stimulate investor interest in STFs, the value added tax should not be imposed on activities supported by STFs.

The state's interests in STFs might be represented by state research and engineering centers. (Chemical research institutes—such as the Karpov Institute—could play the role of engineering firms.) These centers must be able to carry out all preliminary technical and economic examinations of proposed projects and prepare necessary technical documentation for the investors.

Initially the activity of each STF should be oriented toward a specific industry. Over time STFs could become interdisciplinary financial organizations.

Legal establishment of Science and Technology Funds would not be difficult. The main obstacle would be the privileged tax treatment on profits and products.

CONCLUSION

The commercialization of scientific research results is a complex problem. Excellent and prospective scientific results, qualified and experienced personnel, good legislation, a favorable investment climate, and sufficient capital at research institutes are needed to solve the problem. In Russia, excellent research that could be the basis for new technologies abounds. However, qualified managers and good legislation are lacking. The investment climate has begun to change for the better. At the present time, the shortage of capital appears to be the main cause of inefficient commercialization. Without such capital, research institutes cannot develop scientific results to the point at which engineering or industrial companies can clearly understand exactly how to implement a new technology and estimate the profits from doing so. Science and technology funds might be of great benefit in solving this important problem, even in the absence of experienced personnel or proper legislation. Creation of STFs by American partners, industrial and engineering companies, and entrepreneurs would be desirable, as would these groups' more active role in improving the overall investment climate in Russia.

AREAS FOR FURTHER CONSIDERATION

During the final session of the workshop, which was open to the public, the participants suggested steps that should be considered by government agencies and research organizations to assist in the commercialization of technologies. Some of the suggestions were addressed in the papers presented at the workshop, others were based on observations during the field visits in the United States and Russia, and still others emerged during the discussions that followed the formal workshop presentations. There was no effort to reach a consensus on the suggestions, which should be considered to be views of individual workshop participants and not recommendations by the NRC.

Impacts on R&D of Russian Tax and IPR Systems: The Russian government should consider supporting detailed studies of the impacts, both positive and negative, of the current tax regulations and patent system on innovation and technology commercialization as well as the likely impacts of proposed changes. Reports indicate that currently the true tax burden on small enterprises is enormous, often inhibiting the formation and successful development of new firms. In addition, Russian institutes and small businesses reportedly are wary of taking on the burdens of enforcing their intellectual property rights through the courts. Over the past decade small, technology-oriented enterprises have been extremely important in the West for job creation, economic growth, and competitiveness. Studies could concentrate on how additional tax incentives and different approaches to protection of intellectual property could spur innovation in Russia.

Transferring Technology Ownership Rights to Research and Educational Institutions: A joint working group could be established to consider the relevance of the provisions of the Bayh-Dole Act to Russian conditions. This act provides for the transfer of intellectual property rights developed pursuant to the provisions of government funding to nonprofit research and educational institutions. In Russia, the funding agency usually retains the rights for itself, thus reducing incentives among researchers to commercialize innovations.

Industry/University Centers: Russian centers analogous to the U.S. National Science Foundation's Industry/University Cooperative Research Centers should be considered. These centers might be regionally-oriented—for example, located in Nizhnynovgorod, Novosibirsk, Krasnoyarsk, Vladivostok, and Irkutsk as well as Moscow and St. Petersburg—rather than industry-specific. These centers would bring Russian research institutes, universities, and industry together; and American and other foreign universities and industry could be invited to participate as appropriate. In addition to serving as focal points for Russian research and technology, the centers could convene international meetings to consider solutions to barriers to commercialization and international collaboration.

Innovation Incubators: The initial positive experiences at some of the sixteen innovation incubators in Russia should be replicated in other industrial areas. The American experience, such as the experience in the Research Triangle Park region, is of special interest in helping to provide guidance and support for emerging scientific entrepreneurs in Russia. Incubators might be appropriate at locations near closed cities where research institutes are attempting to convert their military-oriented R&D capabilities to provide products and services for civilian markets.

Publicizing Sections of the Russian Tax Code that Impact on R&D: There is considerable confusion among Russian research institutes and enterprises about the tax regulations concerning R&D expenses, income from the use of new inventions, and income of scientific organizations in general. There is even greater confusion as to proposed changes in the tax code. The Russian government should consider ways to clearly communicate to affected parties the existing rules and future regulations as they are enacted, particularly tax incentives for use of innovations and the standards for certification (for tax purposes) of an organization as a scientific entity.

Clarifying Questions of Ownership of Property and Property Rights: Much equipment and many buildings of research institutes have been acquired with federal and local government funding, and the ownership of much of this property has not been settled. In addition, innovations with commercial potential continue to be developed using this property and additional government funding; thus, there remains uncertainty about the rights the federal and regional governments have to these innovations and the circumstances under which these rights can be exercised. Such lingering questions impede the commercialization of innovations, and the Russian government should consider ways to clarify these issues as soon as possible.

Regional Coordination Centers: The Russian government should consider supporting centers with modest budgets to encourage and evaluate technology commercialization activities and related business developments at the regional level. These centers could keep track of available research resources and equipment and facility capabilities, stimulate cooperation among institutes to increase efficient use of limited resources, and facilitate international investment and cooperation in the regions. The experience of state governments in the United States seems particularly relevant to such an approach.

Industrial Consortia, Affiliates, and Related Programs: A number of Russian institutions clearly have the capability to serve as hubs for industrial consortia, affiliate programs, and trade associations in specific technical areas. Institutes should actively pursue opportunities that could lead to sponsored research and valuable contacts with industry. Such programs also could leverage resources to support work that would not be profitable for a single company. As a first step, Russian research institutes should identify technical areas of interest to industry in which the institutes have a comparative R&D advantage. For Russian institutes initiating collective research programs, there are many models for managing intellectual property issues developed by U.S. research consortia which could be adapted to Russian conditions.

Outreach to the Public: To build their customer base, Russian institutes need to better publicize their research, personnel, facilities, and interests. Positive stories about the payoff of both international and domestic projects should be featured. The World-Wide Web is an inexpensive yet expansive medium for disseminating information. For maximum value, institutes should ensure that their home pages are appropriately linked to related pages on the Web. In light of the limitations of the current telecommunications infrastructure in Russia, institutes might seek to establish their home pages on servers in the West.

Utilization of Physical Resources: Russian research institutes should consider how to increase the return on their currently underutilized physical resources. Possibilities include leasing equipment, using space for incubators, and reconfiguring space and equipment for use by consortia or industrial affiliates programs. The availability of such facilities and equipment could be advertised on the World-Wide Web.

Attracting and Retaining Young Scholars: The problems of internal and external brain drain in Russian science and engineering increasingly are apparent, and the overall financial troubles have made it difficult to attract students into technical fields. Institutes should develop programs to expose students to the emerging challenges of science. In addition, they should pursue programs, whenever possible in cooperation with industry, that will help ensure that there are enough scientists and engineers attuned to the needs of the private sector to support the country's future industrial base.

International Linkages: Russian science and technology leaders should explore how western trade associations, professional technology transfer associations (such as the Association of University Technology Managers and the Licensing Executives Society), the Industrial Research Institute, and similar institutions could be models for strengthening internal cooperation and could serve as focal points for cooperation with western counterparts.

Management Training: Russian research organizations could be more proactive in providing management training carefully tailored to the specific needs of their personnel. Training should include technology assessment techniques, preparation of bankable business plans, marketing, and strategic planning. Both scientists and managers should be well-versed in licensing agreements, patenting, and protection of intellectual property rights.

Education in Management Sciences: Few education programs in Russia cover topics directly related to management and commercialization of technology. Several American universities that specialize in this topic could cooperate with Russian institutions to adapt programs to the Russian experience, perhaps working through the Russian network of continuing education programs under the Ministry of Education. In creating such education and training programs, the American and Russian partners should take full advantage of advances in information technology.

Problem-Solving Workshops: Small workshops could be convened among Russian researchers, government officials, and Russian and western industry on a continuing basis to discuss barriers to commercialization. One focus of these workshops could be the tax and patent framework. Topics also might include problems of marketing Russian high-technology products internationally, particularly the importance of international product standards and timing in marketing new products.

APPENDIX A

WORKSHOP ON TECHNOLOGY COMMERCIALIZATION AGENDA

Thursday, March 12, 1998

8:45 a.m.	**Welcome - William A. Wulf, President of the National Academy of Engineering**
9:00 a.m.	**Russian Delegation's Views on Meetings and Visits in North Carolina**
10:00 a.m.	**Presentations on Applied Research Activities in Russia** ● Academician Nikolai Laverov, Russian Academy of Sciences ● Vladimir Meshcheryakov, Russian Agency for Patents and Trademarks
12:00 noon	**Lunch** Speaker: Bardwell Salmon, Reality Wave, Inc. "Commercializing Technology"
1:30 p.m.	**Presentations on Applied Research Experience in the United States of Relevance to Russia** ● Alexander MacLachlan, DuPont Company (retired) ● Richard Dulik, Covington & Burling

- Alexis Clare, New York State College of Ceramics
- Mark Crowell, North Carolina State University

4:00 p.m. **Break**

4:15 p.m. **Additional Russian Presentations**

- Vladimir Litvinenko, Plekhanov Mining Academy
- Alexander Ozerin, Institute of Synthetic Polymer Materials

Friday, March 13, 1998

8:45 a.m. **Additional U.S. Presentations**

- Alvin Trivelpiece, Oak Ridge National Laboratory
- David McNelis, Research Triangle Institute

Additional Russian Presentations

- Sergei Ivantchev, NPO Plastpolimer
- Alexander Simonov, Karpov Physical Chemistry Institute

11:00 a.m. **Work Group Sessions**

- Issues that should be considered by the Russian Government
- Issues that should be considered by Russian R&D groups
- Issues that should be considered for bilateral programs

1:00 p.m. **Lunch**

2:00 p.m. **Final Plenary Session to Consider Work Groups' Discussions and the Final Report**

4:00 p.m. **Adjournment**

APPENDIX B

EXCERPTS FROM THE BAYH-DOLE ACT

§ 200. Policy and objective

It is the policy and objective of the Congress to use the patent system to promote the utilization of inventions arising from federally supported research or development; to encourage maximum participation of small business firms in federally sponsored research and development efforts; to promote collaboration between commercial concerns and nonprofit organizations, including universities; to ensure that inventions made by nonprofit organizations and small business firms are used in a manner to promote free competition and enterprise; to promote the commercialization and public availability of inventions made in the United States by United States industry and labor; to ensure that the Government obtains sufficient rights in federally supported inventions to meet the needs of the Government and protect the public against nonuse or unreasonable use of inventions; and to minimize the costs of administering policies in this area. (Added December 12, 1980, Public Law 96-517, sec. 6(a), 94 Stat. 3019.)

§ 202. Disposition of rights

(a) Each nonprofit organization or small business firm may, within a reasonable time after disclosure as required by paragraph (c)(1) of this section, elect to retain title to any subject invention: *Provided, however,* That a funding agreement may provide otherwise (i) when the contractor is not located in the United States or is subject to the control of a foreign government, (ii) in exceptional circumstances when it is determined by the agency that restriction or elimination of the right to retain title to any subject invention will better promote the policy and objectives of this chapter, (iii) when it is determined by a Government authority which is authorized by statute or Executive order to conduct foreign intelligence or counter-intelligence activities that the restriction or elimination of the right to retain title to any subject invention is necessary to protect the security of such activities, or (iv) when the funding agreement includes the operation of a Government-owned, contractor-operated facility of the Department of Energy primarily dedicated to that Department's naval nuclear propulsion or weapons related programs and all funding agreement limitations under this subparagraph on the contractor's right to elect title to a subject invention are limited to inventions occurring under the above two programs of the Department of Energy. This rights of the nonprofit organization or small business firm shall be subject to the provisions of paragraph (c) of this section and the other provisions of this chapter. (Amended November 8, 1994, Public Law 98-620, sec. 501(3), 98 Stat. 3364.)

(b)(1) The rights of the Government under subsection (a) shall not be exercised by a Federal agency unless it first determines that at least one of the conditions identified in clauses (I) through (iii) of subsection (a) exists. Except in the case of subsection (a)(iii), the agency shall file with the Secretary of Commerce, within thirty days after the award of the applicable funding agreement, a copy of such determination. In the case of a determination under subsection (a)(ii), the statement shall include an analysis justifying the determination. In the case of determinations applicable to funding agreements with small business firms, copies shall also be sent to the Chief Council for Advocacy of the Small Business Administration. If the Secretary of Commerce believes that any individual determination or pattern of determinations is contrary to the policies and objectives of this chapter, the Secretary shall so advise the head of the agency concerned and the Administrator of the Office of Federal Procurement Policy, and recommend corrective actions.

(2) Whenever the Administrator of the Office of Federal Procurement Policy has determined that one or more Federal agencies are utilizing the authority of clause)I) or (ii) of subsection (a) of this section in a manner that is contrary to the policies and objectives of this chapter the Administrator is authorized to issue regulations describing classes of situations in which agencies may not exercise the authorities of those clauses. (Amended November 8, 1994, Public Law 98-620, sec. 501(4A), 98 Stat. 3365.)

(3) At least once each year, the Comptroller General shall transmit a report to the Committees on the Judiciary of the Senate and House of Representatives on the manner in which this chapter is being implemented by the agencies and on such other aspects of Government patent policies and practices with respect to federally funded inventions as the Comptroller General believes appropriate.

(4) If the contractor believes that a determination is contrary to the policies and objectives of this chapter or constitutes an abuse of discretion by the agency, the determination shall be subject to the last paragraph of section 203(2). (Added November 8, 1994, Public Law 98-620, sec. 501(4A), 98 Stat. 3365.)

(c) Each funding agreement with a small business firm or nonprofit organization shall contain appropriate provisions to effectuate the following:

(1) That the contractor disclose each subject invention to the Federal agency within a reasonable time after it becomes known to contractor personnel responsible for the administration of patent matters, and that the Federal Government may receive title to any subject invention not disclosed to it within such time.

(2) That the contractor make a written election within two years after disclosure to the Federal agency (or such additional time as may be approved by the Federal agency) whether the contractor will retain title to a subject invention: *Provide,* That in any case where publication, on sale, or public use, has initiated the one year statutory period in which valid patent protection can still be obtained in the United States, the period for election may be shortened by the Federal agency to a date that is not more than sixty days prior to the end of the statutory period: *And provided further,* That the Federal Government may receive title to any subject invention in which the contractor does not elect to retain rights or fails to elect rights within such times.

(3) That a contractor electing rights in a subject invention agrees to file a patent application prior to any statutory bar date that may occur under this title due to publication, on sale, or public use, and shall thereafter file corresponding patent applications in other countries in which it wishes to retain title within reasonable times, and that the Federal Government may receive title to any subject inventions in the United States or other countries in which the contractor has not filed patent applications on the subject invention within such times.

(4) With respect to any invention in which the contractor elects rights, the Federal agency shall leave a nonexclusive, nontransferable, irrevocable, paid-up license to practice or leave practiced for or on behalf of the United States any subject invention throughout the world: *Provided,* That the funding agreement may provide for such additional rights; including the right to assign or leave assigned foreign patent rights in the subject invention, as are determined by the agency as necessary for meeting the obligations of the United States under any treaty, international agreement, arrangement of cooperation, memorandum of understanding, or similar arrangement, including military agreements relating to weapons development and production. (Amended November 8, 1984, Public law 98-620, sec. 501(5), 98 Stat. 3365.)

(5) The right of the Federal agency to require periodic reporting on the utilization or efforts at obtaining utilization that are being made by the contractor or his licensees or assignees: *Provide* That any such information, as well as any information on utilization or efforts at obtaining utilization obtained as part of a proceeding under section 203 of this chapter shall be treated by the Federal agency as commercial and financial information obtained from a person and privileged and confidential and not subject to

disclosure under section 552 of title 5 of the United States Code. (Amended November 8, 1984, Public Law 98-620, sec. 501(6), 98 Stat. 3365.)

(6) An obligation on the part of the contractor, in the event a United States patent application is filed by or on its behalf or by any assignee of the contractor, to include within the specification of such application and any patent issuing thereon, a statement specifying that the invention was made with Government support and that the Government has certain rights in the invention.

(7) In the case of a nonprofit organization, (A) a prohibition upon the assignment of rights to a subject invention in the United States without the approval of the Federal agency, except where such assignment is made to an organization which has as one of its primary functions the management of inventions (provided that such assignee shall be subject to the same provisions as the contractor); (B) a requirement that the contractor share royalties with the inventor; (C) except with respect to a funding agreement for the operation of a Government-owned-contractor-operated facility, a requirement that the balance of any royalties or income earned by the contractor with respect to subject inventions, after payment of expenses, (including payments to inventors) incidental to the administration of subject inventions, be utilized for the support of scientific research, or education; (D) a requirement that, except where it proves infeasible after a reasonable inquiry, the licensing of subject inventions shall be given to small business firms; and (E) with respect to a funding agreement for the operation of a Government-owned-contractor-operated facility, requirements (i) that after payment of patenting costs, licensing costs, payments to inventors, and other expenses incidental to the administration of subject inventions, 100 percent of the balance of any royalties or income earned and retained by the contractor during any fiscal year, up to an amount equal to five percent of the annual budget of the facility, shall be used by the contractor for scientific research, development, and education consistent with the research and development mission and objectives of the facility, including activities that increase the licensing potential of other inventions of the facility provided that if said balance exceeds five percent of the annual budget of the facility, that 75 percent of such excess shall be paid to the Treasury of the United States and the remaining 25 percent shall be used for the same purposes as described above in this clause (D); and (ii) that, to the extent it provides the most effective technology transfer, the licensing of subject inventions shall be administered by contractor employees on location at the facility. (Amended November 8, 1984, Public law 98-620, sec. 501(7), (8), 98 Stat. 3366.)

(8) The requirements of sections 203 and 204 of this chapter.

(d) If a contractor does not elect to retain title to a subject invention in cases subject to this section, the Federal agency may consider and after consultation with the contractor grant requests for retention of rights by the inventor subject to the provisions of this Act and regulations promulgated hereunder.

(e) In any case when a Federal employee is a coinventor of any invention made under a funding agreement with a nonprofit organization or small business firm, the Federal agency employing such coinventor is authorized to transfer or assign whatever rights it may acquire in the subject invention from its employee to the contractor subject to the conditions set forth in this chapter.

(f)(1) No funding agreement with a small business firm or nonprofit organization shall contain a provision allowing a Federal agency to require the licensing to third parties of inventions owned by the contractor that are not subject inventions unless such provision has been approved by the head of the agency and a written justification has been signed by the head of the agency. Any such provision shall clearly state whether the licensing may be required in connection with the practice of a subject invention, a specifically identified work object, or both. The head of the agency may not delegate the authority to approve provisions or sign justifications required by this paragraph.

(2) A Federal agency shall not require the licensing of third parties under any such provision unless the head of the agency determines that the use of the invention by others is necessary for the practice of a subject invention or for the use of a work object of the funding agreement and that such action is necessary to achieve the practical application of the subject invention or work object. Any such determination shall be on the record after an opportunity for an agency hearing. Any action commenced for judicial review of such determination shall be brought within sixty days after notification of such determination. (Added December 12, 1980, Public Law 96-517, sec. 6(a), 94 Stat. 3020.)

§ 203. March-in rights

(1) With respect to any subject invention in which a small business firm or nonprofit organization has acquired title under this chapter, the Federal agency under whose funding agreement the subject invention was made shall have the right, in accordance with such procedures as are provided in regulations promulgated hereunder to require the contractor, an assignee or exclusive licensee of a subject invention to grant a nonexclusive, partially exclusive, or exclusive license in any field of use to a responsible applicant or applicants, upon terms that are reasonable under the circumstances, and if the contractor, assignee, or exclusive licensee refuses such request, to grant such a license itself, if the Federal agency determines that such-

(a) action is necessary because the contractor or assignee has not taken, or is not expected to take within a reasonable time, effective steps to achieve practical application of the subject invention in such field of use;

(b) action is necessary to alleviate health or safety needs which are not reasonably satisfied by the contractor, assignee, or their licensees;

(c) action is necessary to meet requirements for public use specified by the Federal regulations and such requirements are not reasonably satisfied by the contractor, assignee, or licensees; or

(d) action is necessary because the agreement required by section 204 has not been obtained or waived or because a licensee of the exclusive right to use or sell any subject invention in the United States is in breach of its agreement obtained pursuant to section 204.

(2) A determination pursuant to this section or section 202(b)(4) shall not be subject to the Contract Disputes Act (41 U.S.C. §601 et seq.). An administrative appeals procedure shall be established by regulations promulgated in accordance with section 206. Additionally, any contractor, inventor, assignee, or exclusive licensee adversely affected by a determination under this section may, at any time within sixty days after the determination is issued, file a petition in the United States Claims Court, which shall have jurisdiction to determine the appeal on the record and to affirm, reverse, remand or modify, as appropriate, the determination of the Federal agency. In cases described in paragraphs (a) and (c), the agency's determination shall be held in abeyance pending the exhaustion of appeals or petitions filed under the preceding sentence. (Added December 12, 1980, Public Law 96-517, sec. 6(a), 94 Stat. 3022; Amended November 8, 1984, Public Law 98-620, sec. 501(9), 98 Stat. 3367.)

§ 207. Domestic and foreign protection of federally owned inventions

(a) Each Federal agency is authorized to-

(1) apply for, obtain, and maintain patents or other forms of protection in the United States and in foreign countries on inventions in which the Federal Government owns a right, title, or interest;

(2) grant nonexclusive, exclusive, or partially exclusive licenses under federally owned patent applications, patents, or other forms of protection obtained, royalty-free or for royalties or other consideration, and on such terms and conditions, including the grant to the licensee of the right of enforcement pursuant to the provisions of chapter 29 of this title as determined appropriate in the public interest;

(3) undertake all other suitable and necessary steps to protect and administer rights to federally owned inventions on behalf of the Federal Government either directly or through contract; and

(4) transfer custody and administration, in whole or in part, to another Federal agency, of the right, title, or interest in any federally owned invention.

(b) For the purpose of assuring the effective management of Government-owned inventions, the Secretary of Commerce [is] authorized to-

(1) assist Federal agency efforts to promote the licensing and utilization of Government-owned inventions;

(2) assist Federal agencies in seeking protection and maintaining inventions in foreign countries, including the payment of fees and costs connected therewith; and

(3) consult with and advise Federal agencies as to areas of science and technology research and development with potential for commercial

utilization. (Added December 12, 1980, Public Law 96-517, sec. 6(a), 94 Stat. 3023; Amended November 8, 1984, Public Law 98-620, sec. 501(11), 98 Stat. 3367.)

§ 209. Restrictions on licensing of federally owned inventions .

(a) No Federal agency shall grant any license under a patent or patent application on a federally owned invention unless the person requesting the license has supplied the agency with a plan for development and/or marketing of the invention, except that any such plan may be treated by the Federal agency as commercial and financial information obtained from a person and privileged and confidential and not subject to disclosure under section 552 of title 5 of the United States Code.

(b) A Federal agency shall normally grant the right to use or sell any federally owned invention in the United States only to a licensee that agrees that any products embodying the invention or produced through the use of the invention will be manufactured substantially in the United States.

(c)(1) Each Federal agency may grant exclusive or partially exclusive licenses in any invention covered by a federally owned domestic patent or patent application only if, after public notice and opportunity for filing written objectives, it is determined that

(A) the interests of the Federal Government and the public will best be served by the proposed license, in view of the applicant's intentions, plans, and ability to bring the invention to practical application or otherwise promote the invention's utilization by the public;

(B) the desired practical application has not been achieved, or is not likely expeditiously to be achieved, under any nonexclusive license which has been granted, or which may be granted, on the invention;

(C) exclusive or partially exclusive licensing is a reasonable and necessary incentive to call forth the investment of risk capital and expenditures to bring the invention to practical application or otherwise promote the invention's utilization by the public;

(D) the proposed terms and scope of exclusivity are not greater than reasonably necessary to provide the incentive for bringing the invention to practical application or otherwise promote the invention's utilization by the public.

(2) A Federal agency shall not grant such exclusive or partially exclusive license under paragraph (1) of this subsection if it determines that the grant of such license will tend substantially to lessen competition or result in undue concentration in any section of the country in any line of commerce to which the technology to be licensed relates, or to create or maintain other situations inconsistent with the antitrust laws.

(3) First preference in the exclusive or partially exclusive licensing of federally owned inventions shall go to small business firms submitting plans that are determined by the agency to be within the capabilities of the firms

and equally likely, if executed, to bring the invention to practical application as any plans submitted by applicants that are not small business firms.

(d) After consideration of whether the interests of the Federal Government or United States industry in foreign commerce will be enhanced, any Federal agency may grant exclusive or partially exclusive licenses in any invention covered by a foreign patent application or patent, after public notice and opportunity for filing written objections, except that a Federal agency shall not grant such exclusive or partially exclusive license if it determines that the grant of such license will tend substantially to lessen competition or result in undue concentration in any section of the United States in any line of commerce to which the technology to be licensed relates, or to create or maintain other situations inconsistent with antitrust laws.

(e) The Federal agency shall maintain a record of determinations to grant exclusive or partially exclusive licenses.

(f) Any grant of a license shall contain such terms and conditions as the Federal agency determines appropriate for the protection of the interests of the Federal Government and the public, including provisions for the following:

(1) periodic reporting on the utilization or efforts at obtaining utilization that are being made by the licensee with particular reference to the plan submitted: *Provided,* That any such information may be treated by the Federal agency as commercial and financial information obtained front a person and privileged and confidential and not subject to disclosure under section 552 of title 5 of the United States Code;

(2) the right of the Federal agency to terminate such license in whole or in part if it determines that the licensee is not executing the plan submitted with its request for a license and the licensee cannot otherwise demonstrate to the satisfaction of the Federal agency that it has taken or can be expected to take within a reasonable time, effective steps to achieve practical application of the invention;

(3) the right of the Federal agency to terminate such license in whole or in part if the licensee is in breach of an agreement obtained pursuant to paragraph (b) of this section; and

(4) the right of the Federal agency to terminate the license in whole or in part if the agency determines that such action is necessary to meet requirements for public use specified by Federal regulations issued after the date of the license and such requirements are not reasonably satisfied by the licensee. (Added Dec.12, 1980, Public Law 96-517, sec. 6(a), 94 Stat. 3024.)

§ 212. Disposition of rights in educational awards

No scholarship, fellowship, training grant, or other funding agreement made by a Federal agency primarily to an awardee for educational purposes will contain any provision giving the Federal agency any rights to inventions made by the awardee. (Added Nov. 8, 1984, Public Law 98-620, sec. 501(14), 98 Stat. 3368.)

APPENDIX
C

EXCERPTS FROM THE NATIONAL COMPETITIVENESS TECHNOLOGY TRANSFER ACT OF 1989 WITH THE 1990 AMENDMENTS

Section 3132 Findings and Purpose

(a) FINDINGS -- Congress Finds that --

(1) technology advancement is a key component in the growth the United States industrial economy, and a strong industrial base is an essential element of the security of this country;

(2) there is a need to enhance United States competitiveness in both domestic and international markets;

(3) innovation and the rapid application of commercially valuable technology are assuming a more significant role in near-term marketplace success;

(4) the Federal laboratories and other facilities have outstanding capabilities in a variety of advanced technologies and skilled scientists, engineers, and technicians who could contribute substantially to the posture of the United States industry in international competition;

(5) improved opportunities for cooperation research and development agreements between contractor-managers of certain Federal laboratories and the private sector in the United States, consistent with the program missions at those facilities, particularly the national security functions involved in atomic energy defense activities, would contribute to our national well-being; and

(6) more effective cooperation between those laboratories and the private sector in the United States is required to provide speed and certainty in the technology transfer process.

(b) PURPOSES -- The purposes of this part are to--

(1) enhance United States national security by promoting technology transfer between Government-owned, contractor-operated laboratories and the private sector in the United States; and

(2) enhance collaboration between universities, the private sector, and Government-owned, contractor-operated laboratories in order to foster the development of technologies in areas of significant economic potential.

Section 3133 Authority to Enter into Cooperative Research and Development Agreements

(a) TECHNOLOGY TRANSFER ACTIVITIES - Section 12 of the Stevenson-Wydler Technology Innovation Act of 1980 (15 U.S.C. 3710a) is amended--

(1) in subsection (a) --

(A) by inserting ", and to the extent provided in an agency-approved joint work statement, the director of any of its Government-owned, contractor-operated laboratories" after "Government-operated Federal laboratories";

(B) by striking "Government-owned" and inserting in lieu thereof "(in the case of a Government-owned, contractor-operated laboratory, subject of subsection (c) of this section) for" in paragraph (2); and

(C) by striking "of Federal employees" in paragraph (2);

(2) in subsection (b) --

(A) by inserting ", and to the extent provided in an agency-approved joint work statement, the director of any of its Government-owned, contractor-operated laboratory" after "Government-operated Federal laboratory";

(B) by striking "a Federal" in paragraph (2) and inserting in lieu thereof "a laboratory"; and

(C) by inserting after paragraph (5) the following:

"A Government-owned, contractor-operated laboratory that enters into a cooperative research and development agreement under subsection (a)(1) may use or obligate royalties or other income accruing to such laboratory under such agreement with respect to any invention only (i) for payments to inventors; (ii) for the purposes described in section 14(a)(1)(B)(i), (ii), and (iv); and (iii) for scientific research and development consistent with the research and development mission and objectives of the laboratory.";

(3) in subsection (c)(3)(A), by striking "employee standards of conduct" and inserting in lieu thereof "standards of conduct for its employees";

(4) in subsection (c)(5)(A), by inserting "presented by the director of a Government-operated laboratory" after "any such agreement";

(5) in subsection (c)(5)(B), by inserting "presented by the director of a Government-operated laboratory" after "an agreement presented";

(6) in subsection (c)(5), by adding at the end of the following new subparagraph:

"(C)(i) Any agency which has contracted with a non-Federal entity to operate a laboratory shall review and approve, request specific modifications to, or disapprove a joint work statement that is submitted by the director of such laboratory within 90 days of after such submission. In any case where an agency has requested specific modifications to a joint work statement, the agency shall approve or disapprove any resubmission of such joint work statement within the 30 days after such resubmission, or 90 days after the original submission, which occurs later. No agreement may be entered into by a Government-owned, contractor-operated laboratory under this section before both approval of the agreement under clause (iv) and approval under this clause of a joint work statement.

"(ii) In any case in which an agency which has contracted with a non-Federal entity to operate a laboratory disapproves or requests the modification of a joint work statement submitted under this section, the agency shall promptly transmit a written explanation of such disapproval or modification to the director of the laboratory concerned.

"(iii) An agency which has contracted with a non-Federal entity to operate a laboratory or laboratories shall develop and provide to such laboratory or laboratories one or more model cooperative research and development agreements for the purposes of standardizing practices and procedures, resolving common legal issues, and enabling review of cooperative research and development agreements to be carried out in a routine and prompt manner.

"(iv) An agency which has contracted with a non-Federal entity to operate a laboratory shall review each agreement under this section. Within 30 days after the presentation, by the director of the laboratory, of such agreement, the agency shall, on the basis of such review, approve or request specific modification to such agreement. Such agreement shall not take effect before approval under this clause.

"(v) If an agency fails to complete a review under clause (iv) within the 30-day period specified therein, the agency shall submit to the Congress, within 10 days after the end of that 30-day period, a report in the reasons for such failure. The agency shall, at the end of each successive 30-day period thereafter during which such failure continues, submit to Congress another report on the reasons for the continuing failure. Nothing in this clause relieves the agency if the requirement to complete a review under clause (iv).

"(vi) In any case in which an agency which has contracted with a non-Federal entity to operate a laboratory requests the modification of an agreement presented under this section, the agency shall promptly transmit a written explanation of such modification to the director of the laboratory concerned.";

(7) in subsection (c), by adding at the end the following new paragraph:

"(7)(A) No trade secrets or commercial or financial information that is privileged or confidential, under the meaning of section 552(b)(4) of title 5, United States Code, which is obtained in the conduct of research or as a result of activities under this Act from a non-Federal party participating in cooperative research and development agreement shall be disclosed.

"(B) The director, or in the case of a contractor-operated laboratory, the agency, for a period of up to 5 years after development of information that results from research and development activities conducted under this Act and that would be a trade secret or commercial or financial information that is privileged or confidential if the information had been obtained from a non-Federal party participating in a cooperative research and development agreement, may provide appropriate protections against the dissemination of such information, including exemption from subchapter II of chapter 5, United States Code"; and

(8) in subsection (d)--

(A) by striking "and" at the end of paragraph (1);

(B) by amending paragraph (2) to read as follows:

"(2) the term 'laboratory' means--

"(A) a facility or group of facilities owned, leased, or otherwise used by a Federal agency, a substantial purpose of which is the performance of research, development, or engineering by employees of the Federal Government;

"(B) a group of Government-owned, contractor-operated facilities under a common contract, when a substantial purpose of the contract is the performance of research and development for the Federal Government; and

"(C) a Government-owned, contractor-operated facility that is not under a common contract described in subparagraph (B), and the primary purpose of which is the performance of research and development for the Federal Government,

but such term does not include any facility covered by Executive Order No. 12344, dated February 1, 1982, pertaining to the Naval nuclear propulsion program; and"; and

(C) by adding at the end the following new paragraph:

"(3) the term 'joint work statement' means a proposal prepared for a Federal agency by the director of a Government-owned, contractor-operated laboratory describing the purpose and scope of a proposed cooperative research and development agreement, and assigning the rights and responsibilities among the agency, the laboratory, and any other party or parties to the proposed agreement.".

(b) PRINCIPLES -- Section 12 of the Stevenson-Wydler Technology Innovation Act of 1980 (15 U.SC. 3710a) is amended by adding at the end the following new subsection:

"(g) PRINCIPLES -- In implementing this section, each agency which has contracted with a non-Federal entity to operate a laboratory shall be guided by the following principles:

"(1) The implementation shall advance program missions at the laboratory, including any national security mission;

"(2) Classified information and unclassified sensitive information protected by law, regulation, or Executive order shall be appropriately safeguarded.".

(c) TECHNICAL AMENDMENTS -- Section 14 of the Stevenson-Wydler Technology Innovation Act of 1980 (15 U.SC. 3710a) is amended --

(1) in subsection (a)(1) by inserting "by Government-operated Federal laboratories" after "entered into"; and by striking "11" and inserting in lieu thereof "12";

(2) in subsection (a)(1)(B)(ii), by inserting ",including payments to inventors and developers of sensitive or classified technology, regardless of whether the technology has commercial applications" after "that laboratory"; and

(3) in subsection (a)(1)(B)(iv), by striking "Government-operated".

(d) CONTRACT PROVISIONS -- (1) Note later than 150 days after the date of enactment of this Act, each agency which has contracted with a non-Federal entity to operate a Government-owned laboratory shall propose for inclusion in that laboratory's operating contract, to the extent not already included and subject to paragraph (6), appropriate contract provisions that--

(A) establish technology transfer, including cooperative research and development agreements, as a mission for the laboratory under section 11(a)(1) of the Stevenson-Wydler Technology Innovation Act of 1980;

(B) describe the respective obligations and responsibilities of the agency and the laboratory with respect to this part and section 12 of the Stevenson-Wydler Technology Innovation Act of 1980;

(C) require that, except as provided in paragraph (2), no employee of the laboratory shall have a substantial role (including an advisory role) in the preparation, negotiation, or approval of a cooperative research and development agreement if, to such employee's knowledge--

(i) such employee, or the spouse, child, parent, sibling, or partner of such employee, or an organization (other than the laboratory) in which such employee serves as an officer, director, trustee, partner, or employee

(I) holds a financial interest in any entity, other than the laboratory, that has a substantial interest in the preparation, negotiation, or approval of the cooperative research and development agreement; or

(II) receives a gift or gratuity from any entity, other than the laboratory, that has a substantial interest in the preparation, negotiation, or approval of the cooperative research and development agreement; or

(ii) a financial interest in any entity, other than the laboratory, that has a substantial interest in the preparation, negotiation, or approval of the cooperative research and development agreement, is held by any person or organization with whom such employee is negotiating or has any arrangement concerning prospective employment;

(D) require that each employee of the laboratory who negotiates or approves a cooperative research and development agreement shall certify to the agency that the circumstances described in subparagraph (C)(i) and (II) do not apply to such employee;

(E) require the laboratory to widely disseminate information on opportunities to participate with the laboratory in technology transfer, including cooperative research and development agreements; and

(F) provides for an accounting of all royalty or other income received under cooperative research and development agreements.

(2) The requirements described in paragraph (1)(C) and (D) shall not apply in a case where the negotiating or approving employee advises the agency that reviewed the applicable joint work statement under section (c)(5)(C)(i) of the Stevenson-Wydler Technology Innovation Act of 1980 in advance of the matter in which he is to participate and the nature of any financial interest described in paragraph (1)(C), and where the agency employee determines that such financial interest is not so substantial as to be considered likely to affect the integrity of the laboratory employee's service in that matter.

(3) Not later than 180 days after the date of enactment of this Act, each agency which has contracted with a non-Federal entity to operate a Government-owned laboratory shall submit a report to Congress which includes a copy of each contract provision amended pursuant to this subsection.

(4) No Government-owned, contractor-operated laboratory may enter into a cooperative research and development agreement under section 12 of the Stevenson-Wydler Technology Innovation Act of 1980 unless--

(A) that laboratory's operating contract contains the provisions described in paragraph (1)(A) through (F); or

(B) such laboratory agrees in separate writing to be bound by the provisions described in paragraph (1)(A) through (F).

(5) Any contract for a Government-owned, contractor-operated laboratory entered into after the expiration of 150 days after the date of enactment of this Act shall contain the provisions described in paragraph (1)(A) through (F).

(6) Contract provisions referred to in paragraph (1) shall include only such provisions as are necessary to carry out paragraphs (1) and (2) of this subsection.

Appendix D

COMMERCIALIZING TECHNOLOGY

On March 12, Bardwell Salmon gave a broad overview of the key components of technology commercialization from his perspective as an investor and entrepreneur. Mr. Salmon is chairman of Technology Capital Network, a New England regional network for investors and entrepreneurs located at the Massachusetts Institute of Technology, and chairman of RealityWave, Inc., a small company focusing on three-dimensional internet technology.

According to Mr. Salmon, there are four key ingredients for success: people, technology, markets, and money. **People** need an environment that provides an incentive to succeed, coupled with some personal loss for failure, and the freedom to excel. To increase the chance of success, individuals require education in entrepreneurship and mentors. Finally, the small technology-oriented businesses need a source of talent for workers and advisors, and so locating close to a university can boost the probability of successful commercialization.

The second important factor is having a **technology** which people will demand. The innovation must be cutting-edge and difficult for others to duplicate. In order to protect the technology, there must be an adequate legal infrastructure, including a well-developed patent system.

The other two areas, markets and money, are the ones in which innovators often have the most trouble. Previously, **markets** were thought of in geographic terms, with local, national, and global markets representing a series of concentric circles. Now new technologies may have very small niche markets, and the customers may be scattered across the globe. For this reason, those interested in commercializing technology must think early about the complex issues of distribution and protection of intellectual property rights that accompany working in a global market.

Finally, start up companies need **money**. In the United States, there is a wide variety of sources for funds to commercialize technology, including risk capital and individuals, who are known as "angels." Venture capitalists are critical to many innovative companies but they are actually involved in a small number of cases, and one should not overemphasize their importance to technology commercialization. Many small firms obtain a large boost from alliances with larger corporations, which increasingly work as "hunters and gatherers" of new technology instead of owning and developing it themselves. Finally, beyond the existence of potential financing, governments must ensure that there are tax and other incentives for investors and entrepreneurs to risk their money in technology commercialization.

APPENDIX E

U.S. PATENT LAW PROVISIONS THAT PROMOTE UNIVERSITY-BASED PATENTING AND TECHNOLOGY TRANSFER

Kenneth D. Sibley
Myers Bigel Sibley & Sajovec, P.A.

The United States currently operates under the Patent Act of 1952, which was codified as Title 35 of the United States Code. A number of changes have been made to the patent statute since 1952, but the basic framework remains that specified by the original act.

The university community had little input to the 1952 Act. The major change affecting university patents has been the passage and implementation of the Bayh-Dole Act in 1982. Other than the Bayh-Dole Act, however, the University community has had little input into revisions to the patent statute.[1] Clearly, it has had nowhere near the influence on statutory changes that various segments of industry have had.

The patent statute is interpreted by case law, most notably the case law promulgated by the U.S. Court of Appeals for the Federal Circuit (the "CAFC"). The CAFC was created in 1982 to strengthen the patent system in the United States by providing a single appeals court for all patent cases. Although the CAFC has carried out this task admirably, it was not created to provide any special benefit to the university community, and special considerations of universities have had little or no impact on the CAFC's decisions.[2]

---

[1] The Association of University Technology Managers is among the groups that advocates legislative change on behalf of the university community.

[2] Indeed, several cases reject any special consideration to the unique problems posed by university research. For example, see *Griffith v. Kanamaru,* 816 F.2d 624 (Fed. Cir. 1987).

None of the foregoing is meant to suggest that the U.S. patent statute contains no provisions that are beneficial to universities. Several provisions that the university community can use to its advantage are listed and discussed below. However, U.S. patent statute has not been specially adapted to benefit the university community. Provisions in the patent laws of other jurisdictions also can be of benefit to a research university.

THE GRACE PERIOD

The United States is a grace-period jurisdiction: if a professor publishes his or her invention, he or she has one year from the date of first publication to file a patent application before that invention passes into the public domain.[3] If the professor waits more than one year, then patent rights are forfeited. Most other jurisdictions are **absolute novelty** jurisdictions. In an absolute novelty jurisdiction, a patent application must be placed on file before the invention is published (or, in many jurisdictions, disclosed to the public in any way, such as by a speech) or rights to that invention are lost.

The advantages of the grace period to the university community are great. Industrial research can be kept secret for long periods of time. University research, in contrast, is made known to the public much more rapidly. Because industry has more time to develop a line of research before it is published, it also has more time to decide whether a particular invention is worth the cost of filing a patent application. Universities, on the other hand, have comparably small budgets with which to pursue patent filings and simply cannot afford to file a patent application on every new invention that is published. Exacerbating the problem is that university research, while ground-breaking and innovative, may not be motivated by an immediate commercial objective. Without a commercial objective (and corresponding market information), deterring whether a particular invention, however worthy, should have patent protection is difficult.

Whenever possible, American universities attempt to preserve potential patent rights in jurisdictions outside the United States. The U.S. grace period is of no consequence in preserving such rights, and if preserving patent rights in jurisdictions outside the United States is critical to securing business interest, U.S. universities must operate under an absolute novelty rule.[4] Nevertheless, when a commitment to a patent filing simply cannot be made before a disclosure of the invention and causes the loss of patent rights in absolute novelty jurisdictions, the one-year grace period provides time to determine (through

[3] *See* 35 USC §102(b).

[4] The new provisional filing system is thought to provide a quick, inexpensive means to preserve patent rights outside the United States. For reasons that are beyond the scope of this paper, this belief is incorrect.

peer review, industry comment, and the like) whether the invention should be pursued in at least a U.S. filing.

AN OPEN DOOR TO PATENTABLE SUBJECT MATTER

A broad variety of different items and processes can be patented in the United States. In addition to compounds, compositions, machines, products, and methods of making and using the same, diagnostic procedures, methods of treatment with known compounds, microorganisms, plants (including transgenic plants and new, classically bred plants), animals, biological process inventions, computer hardware, and computer software can all be patented. Indeed, a famous quote from the U.S. Supreme Court is that the patent statute is intended to protect "anything under the sun that is made by man."[5]

An established industry can adapt to the statutory subject matter requirement: pharmaceutical industries can bias their discovery process in favor of new compounds, medical device companies can bias their discovery process in favor of new apparatus rather than new procedures, computer software companies can bias their discovery process in favor of copyright protection for detailed programs rather than broad concepts, and so forth. However, for the university research community, some of the most attractive areas for patent protection are groundbreaking technologies, or "platform" technologies, that can lead to the development of a new industry (or at least substantially modify an old industry). Stated otherwise, the university community is often in the position of pursuing patents on "things" that have not been the subject of extensive patent activity in the past. Protection of this type of technology is much easier when the patent statute, case law, and judicial system maintain an open door to patentable subject matter.

GENERIC PROTECTION IS AVAILABLE

Reasonably generic protection can be obtained for most inventions in the United States: that is, the patentee cannot only preclude others from making, using, or selling that which the patentee has made, but can preclude others from making, using, or selling logical and reasonable extensions of that which the patentee has made. For example, the patentee may develop a new compound containing a fluorine atom, where any halogen atom may be acceptable in place of the fluorine atom. The patent applicant can extend the claims of the patent, and hence the coverage of the patent, to these alternate versions of the invention, without being required to synthesize every conceivable compound covered by the claim. Indeed, "generic" patent claims commonly cover thousands, and even

[5] *Diamond v. Chakrabarty,* 447 U.S. 303 (1980).

millions, of different permutations and variations on the invention (or different "embodiments") with only ten or twenty examples of things actually made.[6]

The university-based inventor is usually more interested in making a "proof of principle" rather than making repeated demonstrations that different variations of the invention worked. Indeed, even if the university-based inventor is willing to test different variations of an invention to help support a broad patent position, the funding for this type of research is often not available. Hence, the availability of generic patent protection under U.S. law is a great advantage to university inventors.

THE LAW OF INVENTORSHIP

The United States is a "first-to-invent" jurisdiction, whereas many other jurisdictions are "first-to-file" jurisdictions. That is, one who failed to file a patent application before another, independent, inventor, can nevertheless prove that he or she was the first to make the invention, and hence the one entitled to the patent, through an "interference" proceeding in the U.S. Patent and Trademark Office.

Given the pressure to publish early, the frequent need to preserve patent protection in "first-to-file" jurisdictions outside the United States, and the high cost of interference proceedings, the mere fact that the United States is a first-to-invent jurisdiction is not of great advantage to the university community. However, a side benefit of the first-to-invent system is that the United States has a well-developed law of inventorship. In general, the law of inventorship credits to the person or persons who contribute to the "conception" of an invention, as defined by any claim in a patent application. Conception is the mental portion of the inventive act, rather than the physical portion (called the "reduction to practice"). Hence, an inventor can communicate his invention to others, who may then reduce the invention to practice without depriving the inventor of his status as the true inventor. This rule should be of great benefit and protection to the university research community, where new ideas are generated and exchanged at a rapid pace.

Unfortunately, the law of inventorship in the United States does not always protect the university inventor as it should. There is frequent confusion over the law and difficulty in documenting the specific origin of ideas. Ideas that should be credited to a university-based inventor are found in the patent filings of other parties outside the university more often than they should. Universities are constantly involved in studying the patent filings of others to ensure that a inventorship has been properly attributed. They at least have an established, favorable body of law on this point to which they can refer.

---

[6] The scope of protection is limited by the "enablement" requirement set forth in 35 USC §112.

TREATY MEMBERSHIPS

The United States, like the Russian Federation, is party to a number of treaties that help defer the costs of filing foreign patents. These treaties include the Paris Convention and the Patent Cooperation Treaty. Careful use of these treaties allows the university to preserve patent rights in other jurisdictions for 30 months beyond the original filing while efforts are made to locate a licensee willing to support the cost of filing for foreign patents. Without a licensee willing to support such filings, the high costs of extraterritorial prosecution on a speculative patent filing are impractical for a university to cover in all but the most extraordinary of cases.

THE PROFESSIONALISM OF THE UNIVERSITY TECHNOLOGY TRANSFER COMMUNITY

Many stories circulate about enormous patent costs of university inventions that, while based on outstanding science, never led to a successful license. In general, these stories relate back to the early days of university technology transfer, shortly after passage of the Bayh-Dole act. Few patents ultimately cover commercial products and most patent applications must be filed before their commercial viability is confirmed. However, it has become apparent that university technology transfer programs cannot afford to pursue patent protection on highly speculative or extremely early-stage technology, no matter how meritorious the underlying science may be. These are extremely difficult judgments to make, but they are judgments that university-based technology transfer specialists must deal with every day. Through the efforts of professional organizations such as the Association of University Technology Managers (AUTM), the university community has become (in relatively few years) much more sophisticated at making these types of judgments. In short, they have brought industrial patent strategy considerations into the world of university technology transfer. Without such efforts, the university technology transfer program would quickly sink under the weight of managing patent portfolios that have no reasonable business prospects. This author applauds the professionals who carry out this task, wherever they are located.

APPENDIX F

DESCRIPTION OF THE CENTENNIAL CAMPUS

Slated for development over the next twenty years, the Centennial Campus of the North Carolina State University will consist of a dozen or more research clusters made up of university, corporate, and government laboratories, a hotel-conference center, retail stores, and housing situated around a central lake. The first cluster, on the north side of Lake Raleigh, is already alive with activity.

Nine major buildings are now completed or in planning/ construction, based on a forward-looking infrastructure of telecommunications highways, utilities, roads, parking, stormwater system, and sanitary sewers.

Research Building I: Houses University engineering and physical sciences centers of excellence.

Research Building II: Houses corporate partners, University physical science research labs, and engineering centers of excellence.

Research Building III: Houses the National Weather Service, corporate partners, and earth and atmospheric researchers.

Research Building IV: A 70,000-square-foot, multi-tenant building for university and corporate partners in engineering, transportation, energy and environmental research.

College of Textiles Complex: Teaching, research, outreach, and administrative facilities, as well as the Model Manufacturing Center.

Corporate Research Center I: Houses the U.S. Headquarters for Power Transmission and Distribution for multinational ABB Asea Brown Boveri.

Partners Building I: An 80,000-square-foot multi-tenant building for industry and government partners in biotechnology and environmental research.

Partners Building II: A 60,000-square-foot facility to accommodate industrial and computer engineering partners.

The Engineering Graduate Research Center: A 132,000-square-foot advanced laboratory facility to house researchers in civil, electrical, mechanical, materials, computer, and software engineering.

Also in the planning stages are new clusters surrounding university anchor facilities for advanced communications technologies, biotechnology, environmental sciences, and pre-college educational outreach.

Source: North Carolina University Centennial Campus, 1995

APPENDIX G

INNOVATION RESEARCH FUND

The Innovation Research Fund (the "IRF") is a revolving venture capital program of the North Carolina Technological Development Authority, Inc. (the "TDA"). The IRF provides flexible financing to emerging innovation-oriented business in North Carolina. Since its formation in 1984, the IRF has provided venture capital to about 70 companies ranging from agriculture and biotechnology to computer software.

Typical IRF investments range from $50,000 to $250,000 per company. Financing, usually in the form of equity, is provided to companies with high-growth, job-creation potential. Funds are normally stage into the business, based upon company needs and performance milestones. The IRF prefers to invest with other firms. However, in some cases, the Innovation Research Fund may lead the investment process.

In addition to funding, the IRF provides assistance to its portfolio companies through active participation in the company's direction. While the IRF does not become directly involved in the day-to-day management of portfolio companies, the Fund can help portfolio companies find corporate partners, new employees, and sources of additional funding. These services are designed to help the company meet its operating goals, and to allow the Fund to monitor the company's progress.

The IRF investment approach is to focus on early-stage, high-growth businesses where there is an opportunity for significant long-term capital appreciation and job creation in North Carolina. The IRF concentrates on proposals with outstanding management teams and high-quality products in expanding markets. Investments are made throughout North Carolina, in many industries and in any product or serve that promises an annualized return exceeding 50 percent.

New investment opportunities are reviewed on a continuous basis. In order to be considered for funding, entrepreneurs should present a complete business plan which includes:

- Background of management team
- Description of product or service
- Market opportunity and direction
- Competitive analysis
- Current financial statements
- Financial plan with projected growth through the next five years
- Amount of funding requested
- Use-of-funds description

Companies seeking funding are advised to start the process several months before funds are needed. The IRF will evaluate the strength of the business and if, deciding to pursue the opportunity, will perform due diligence and meet with management to reach an investment agreement.

Source: North Carolina Technological Development Authority, Inc.

APPENDIX H

FIRST FLIGHT VENTURE CENTER: A BUSINESS INCUBATOR IN RESEARCH TRIANGLE PARK

The First Flight Venture Center (FFVC) is a business incubator serving the initial location needs of research-based entrepreneurs. The Center's mission is to increase the number of successful technology-based small companies originating in or relocating to the Research Triangle Park region of North Carolina. The Center provides office and lab facilities, along with business equipment and information services, to help accelerate the investigation and validation of innovative technical and commercial concepts by early stage ventures. By expediting the demonstration of technical and commercial feasibility, FFVC members are positioned to compete for the management and capital resources required for growth.

The FFVC is a 28,500 square foot facility available for short-term leasing. The Center's offices and wet labs range for 115 to 345 square feet. Flex-Spaces up to 1,000 square feet, suitable for proto-type development and manufacturing are also available. The facility can accommodate approximately 20 early-stage companies engaged in a diversity of research and product development efforts.

Membership in the Center is open to both tenant and non-tenant research-based entrepreneurial companies. Members have access to the Center's shared common area, conference rooms, classrooms, AV equipment and business services (receptionist, phone answering, postage metering, fax, copier, etc.) on a usage fee basis. Members are also entitled to attend seminars, workshops and informal gatherings held at the Center, and they are eligible to participate in any Center-sponsored programs.

The Center is managed by the North Carolina Technical Development Authority, Inc. (NCTDA), a non-profit corporation established in 1983. Since its inception, NCTDA, has assisted the growth of technology-based entrepreneurial companies through early-stage equity investment in more than 60 new ventures and through the establishment of 23 business incubators across North Carolina.

Source: North Carolina Technological Development Authority, Inc.

APPENDIX I

NIST ADVANCED TECHNOLOGY PROGRAM

About ATP

The National Institute of Standards and Technology (NIST) Advanced Technology Program (ATP) is a unique partnership between government and private industry to accelerate the development of high-risk technologies that promise significant commercial payoffs and widespread benefits to the economy. The ATP has several critical features that set it apart from other government R&D programs:

- The goal of the ATP is economic growth and the good jobs and quality of life that come with economic growth--opening new opportunities for U.S. business and industry in the world's markets by fostering enabling technologies that lead to new, innovative products, services, and industrial processes. For this reason, ATP projects focus on the technology needs of U.S. industry, not those of government. The ATP is industry driven, which keeps the program grounded in real-world needs. Research priorities for the ATP are set by industry: for-profit companies conceive, propose, co-fund, and execute ATP projects and programs based on their understanding of the marketplace and research opportunities.
- The ATP does not fund product development. It supports enabling technologies that are essential to the development of new products, processes, and services across diverse application areas. Private industry bears the costs of product development, production, marketing, sales, and distribution,
- ATP awards are made strictly on the basis of rigorous peer-reviewed competitions designed to select the proposals that are best qualified in terms of the technological ideas, the potential economic benefits to the nation (not just the applicant), and the strength of the plan for eventual commercialization of the results. Expert reviewers (without conflict of interest) drawn from the business community, government, and academia

carefully examine and rate each proposal according to published selection criteria that focus on both business and technical potential.

- The ATP has strict cost-sharing rules. Joint ventures must pay at least half of the project costs. Single companies working on ATP projects must pay all indirect costs associated with the project. (This provision encourages small companies, particularly start-ups, that often have much lower overhead rates than large firms).
- ATP support does not become a perpetual subsidy or entitlement--each project has goals, specific funding allocations, and completion dates established at the outset. Projects are monitored and can be terminated for cause before completion.
- The ATP benefits companies of all sizes. ATP funding stimulates companies of all sizes to take on greater technical challenges with larger, broader, and faster payoff potential for the nation--benefits that extend well beyond the innovators--than they could or would do alone. For smaller start-up firms, early support from ATP can spell the difference between success and failure. To date, over half (53 percent) of the ATP awards have gone to individual small businesses or to joint ventures led by a small business. Large firms can work with the ATP, especially in joint ventures, to develop critical, high-risk technologies that would be difficult for any one company to justify because, for example, the benefits spread across the industry as a whole. Universities and non-profit independent research organizations also play significant roles as participants in ATP projects. More than 100 different universities are involved in more than 180 ATP projects as either joint-venture participants or subcontractors.
- Since its inception, the ATP has made economic evaluation of the outcomes of ATP projects a central element of its operations. The ATP has developed and implemented a thorough measurement program that pushes the state of the art in evaluating the long-term outcomes of R&D investment.

Competitions

Until 1994, the ATP used general competitions open to proposals in all areas of technology as its sole investment mechanism. Since then, the ATP has added another element to its investment strategy--focused program competitions. Each type of competition has its unique advantages. General competitions ensure that all good ideas receive consideration, no matter what the technology area. Focused programs create a mechanism to provide critical-mass support for high-risk, enabling technologies in particular technology areas identified by U.S. industry as offering especially important opportunities for economic growth.

An ATP focused program identifies a specific set of research and business goals that require the parallel development of a suite of interlocking R&D projects. By managing groups of projects that complement and reinforce each other, the ATP reaps the benefits of synergy and, in the long run, can have a stronger impact of U.S. technology and the economy.

Focused programs are developed in response to specific suggestions received from industry and academia. In the form of white papers, the proposals outline a specific technology area and describe the potential for U.S. economic benefit, the technical ideas available to be exploited, the strength of industry commitment to the work, and the reasons why ATP funding is necessary to achieve well-defined research and business goals.

Areas that attract particularly strong interest--30 or more white papers from different sources proposing the same general effort are not unusual--then are developed further through discussions with industry, meetings, workshops, and other interactions.

Within a focused program, the ATP holds special competitions open only to project proposals that would advance the goals of the specific program. Specific projects are selected through the normal ATP competitive review process. The ATP has received over 1,000 white papers suggesting specific focused program areas. Drawing from more than 300 of these, the ATP has established 17 focused programs to date.

Source:ATP Overview, National Institute of Standards and Technology, February 1998.

APPENDIX J

THE INDUSTRIAL RESEARCH INSTITUTE, INC.

IRI Mission —
To Enhance the Effectiveness of Technology Innovation in Industry

Purposes of the Institute

1. Identify and promote effective techniques for the organization and management of research, development, and engineering in support of technological innovation.
2. Encourage high standards in technological innovation.
3. Develop methods for determining the effectiveness of technological innovation, and promote an understanding of the value of technological innovation to the economy, industry, and society.
4. Strengthen understanding of business issues by technology leaders as well as business leaders' understanding of the technological innovation process.
5. Monitor and clarify government policy issues that relate to technological innovation, act as an effective source of information to the U.S. government, and afford member companies opportunities to influence policies.
6. Foster cooperation on a worldwide basis with academia, government, and other organizations active in technological innovation.
7. Provide member-company representatives a forum for building a network of contacts among their peers.

Industry recognizes that research and development are indispensable to the security and progress of a nation. Concern for improvement of the environment, conservation of resources, and a better life for all persons underlie the importance of research. A 130 billion dollar enterprise, industrial research and development programs in the United States utilize the services of well over one-half million scientists and engineers.

Organization and management of these efforts have given rise to mutual problems, such as selection, control, and termination of projects and evaluation of their results; barriers to innovation; professional development of research personnel; effect of societal influences on R&D; and many others.

Founded in 1938 under the auspices of the National Research Council, the Industrial Research Institute (IRI) is an association of over 265 companies that provides a means for the coordinated study of problems confronting managers of industrial research and development. IRI was incorporated in the State of New York on April 17, 1945, as a non-profit, 501(c)(6) organization.

Activities of the Industrial Research Institute cover current as well as long-range problems in the management of research and technology. Primarily educational and informational, these activities include semi-annual meetings, seminars, study groups, a bi-monthly journal Research-Technology Management, and periodic newsletters.

Procedures for carrying out the Purposes of the Institute are subject to approval of a Board of Directors. Plans, policies, and proposals evolve from a number of standing committees, and are implemented by a headquarters staff of ten persons plus two editorial consultants.

Membership in the Institute is taken in the name of the company upon the payment of annual dues of $3,200. A requirement for membership is that the company maintain a technical staff and laboratory for industrial research in the United States, Canada, or Mexico, and shall either be engaged primarily in industrial production or be the research subsidiary of a company so engaged, or shall be a service company whose industrial product is technical service, information technology, or software. Institute members are located mainly in the United States, although several are in Canada, Europe, and Japan. Industries represented by these companies include chemical, metal-producing, paper, textile, pharmaceutical, food, petroleum, electronics, computers, rubber, aerospace, transportation, and R&D services. The policy on admission to IRI provides for a moderate annual increase in membership while avoiding undue predominance of companies in any one industrial classification.

Source: Industrial Research Institute, Inc., 1998

APPENDIX K

NATIONAL SCIENCE FOUNDATION INDUSTRY/UNIVERSITY COOPERATIVE RESEARCH CENTERS PROGRAM

Introduction

The Industry/University Cooperative Research Centers (I/UCRC) Program was initiated in 1973 to develop long term partnerships among industry, academe, and government. The National Science Foundation invests in these partnerships to promote research programs of mutual interest, contribute to the Nation's research infrastructure base, and enhance the intellectual capacity of the engineering workforce through the integration of research and education.

The Centers are catalyzed by a small investment from NSF and are primarily supported by Center members, with NSF taking a supporting role in their development and evolution. The I/UCRC Program offers five-year awards to Centers that meet the I/UCRC Program requirements. This five-year period allows for the development of a strong partnership between the academic researchers and their industrial and government members. After five years, Centers that continue to meet the I/UCRC Program requirements may apply for a second five-year award. These awards allow Centers to continue to grow and diversify their industrial membership. After ten years, the Centers are expected to be fully supported by industrial, other Federal agency, and state and local government partners.

Defining Characteristics of an I/UCRC

A Center in the I/UCRC Program:

- develops a partnership among academe, industry, and other organizations participating in the center;

- consults with Center members to set a research agenda focused on shared research interests and opportunities;
- shares the intellectual property developed by the Center equally among Center members;
- has Center members monitor and advise on the progress of the research, which speeds two-way transfer of knowledge between universities and industry;
- has industrial and other partners that are the primary financial resource for the Center;
- has a formal structure and policies for Center members outlined in an I/UCRC membership agreement;
- relies primarily on graduate student involvement in the research projects, thus developing students who are knowledgeable in industrially relevant research;
- has a Center Director, based at a university or college, who is responsible for all Center activities; and
- has formal evaluations of the partnership conducted by an independent evaluator.

Eligibility

Universities and colleges with sufficient research and graduate education capabilities are eligible as lead institutions for I/UCRC Program support. Since a comprehensive range of disciplines and skills is necessary to address the research issues of interest to industry, a critical mass of interdisciplinary research capabilities is required to form a Center. In order to ensure a sufficiently broad base of research expertise, multiple universities or colleges are encouraged to partner in forming a Center. Each partner university or college is expected to attract industrial support to the Center.

I/UCRC Program Operational Requirements

Non-NSF Support

To be eligible for the I/UCRC Program, a Center is required to obtain a total of at least $300,000 annually in cash membership fees from a minimum of six Center members. This is the minimum funding needed to support a vital research agenda and to ensure the Center can support a number of students and research projects. The minimum number of members required produces a critical mass of partners and encourages a more generic research program. In general, Center members are industrial firms, although some may be other organizations such as Federal agencies. A Center may designate a number of membership categories with varying levels of membership fees and member benefits. However, there must be at least one membership category with membership fees of $25,000 or higher per year with at least three members participating in the Center at that level. Other membership categories with lower fees may be designated to encourage small company participation in the Center.

University cost sharing is required for a Center in the I/UCRC Program.

Center Policies and Management

A Center differs from a group of researchers performing collaborative research in that it has a formal structure that encompasses a substantial number of projects, several investigators and a group of students whose research is part of the Center. For Centers in the I/UCRC Program, this structure includes a management organization and policies that are outlined in a membership agreement signed by all Center members. The membership agreement delineates policies dealing with intellectual property rights, publication delays, membership fees and rights, university cost sharing, etc. A sample membership agreement, which may be used as a guideline, is available on the I/UCRC web site at <http://www.eng.nsf.gov/eec/i-ucrc.htm>.

In order to integrate the research skills and desires of Center faculty and the research needs of the Center members, a successful Center in the I/UCRC Program has the following management structure:

- A Director who is responsible for all aspects of Center operation. The Center Director is the NSF Principal Investigator (PI) and has primary responsibility for administering the award in accordance with NSF's Grant General Conditions (GC-1) and the I/UCRC Program.
- An Industrial Advisory Board (IAB) that reviews ongoing and completed activities and selects new projects.
- A University Policy Committee that facilitates the operation of the Center within the university or universities to help assure recognition for participation in the Center in tenure and promotion decisions, and to assure that the research is appropriate for graduate education.

NSF Oversight and Evaluation

Operating Centers are required to submit a short annual report on progress and plans 90 days prior to the anniversary of their NSF award date. The report, which will be used as a basis for continued I/UCRC Program support, will include:

- major accomplishments for the Center's most recently completed fiscal year (i.e. scientific and technological developments and significant technology transferred to members);
- research goals for the current year;
- a short description of the processes used to interact and communicate with Center members (i.e. the project selection process used by the Center, reports generated, etc.);

- quantitative information from the most recently completed fiscal year such as number of students, faculty, and industrial members involved in the Center, degrees granted to students involved in Center activities, amounts and sources of income to the Center, and lists of patents, licenses and publications created;
- NSF budget forms, statement of fund obligation and statement of university cost-sharing;
- a copy of the Center's membership agreement; and
- a certification of the receipt of annual cash membership fees signed by the Center or Site Director and an official from the Sponsored Research Office.

NSF requires that the industry/university interaction of each Center be independently observed and evaluated during its operational phases by an independent evaluator, who is usually chosen from within the university but not from the department receiving Center funding. This gives both NSF and the Center's management feedback on the health and evolution of the partnership between Center researchers and members.

The Center evaluator is responsible for:

- preparing an annual review of Center activities with respect to industrial collaboration during the
- previous year (which is appended to the Center's annual report to NSF);
- Conducting a survey (using an instrument common to all Centers) of all Center participants to probe the participant satisfaction with Center activities;
- compiling a set of quantitative indicators to analyze the management and operation of the Center; participating in the IAB and any other relevant meetings;
- performing exit interviews to determine why departing companies chose to withdraw from the Center; and

feeding information on the quality of the industry/university partnership back to the Center for continuous improvement.

Concept Paper

A concept paper describing the proposed Center must be submitted to NSF for internal review. The concept paper must be approved by an I/UCRC Program Director before a proposal for either a planning grant or operational Center award will be accepted. Approval decisions will be made periodically, but no later than three months after receipt of a concept paper. The proposed Centers that fit within the industry/university collaborative scope are considered potentially viable, and do not significantly duplicate the research focus of other Centers funded in the program will be encouraged to submit a proposal for a planning grant. Those Centers that are ready for full operation under the requirements of the I/UCRC Program may submit a proposal for an operational Center award.

Planning Grant Proposal

A planning grant supplies funds to study the feasibility of developing the industry/university interaction necessary to establish and support a Center. The funds are used to bring together potential members to establish a research plan that fits their needs. Planning grant proposals may be reviewed internally or through external peer review, and awards do not exceed $10,000.

Operational Center Awards

Operational Centers may be based at a single university or college, or may be initiated by more than one institution. The initial I/UCRC award to a Center has a potential duration of five years, assuming sufficiently meritorious achievement and success at maintaining leverage of NSF support. I/UCRC Program support shall be up to $100,000 annually for the duration of the initial award to enable the Center to manage its proposed research and education program effectively in partnership with its other sponsors. NSF support is intended to augment the support the Center receives from industry and other sponsors. Proposals for operational Center awards are evaluated using external peer review.

Some funds are available to support institutions joining existing Centers in the I/UCRC Program. For multi-university Centers, additional funds are available to the lead institution to offset the added administrative burden.

The initial I/UCRC award may be extended for an additional period of up to five years following a successful renewal review guided by peer evaluation and favorable recommendation by the NSF Program Director. NSF I/UCRC Program support for the second five-year award shall be up to $50,000 annually.

Competition to Start New I/UCRC Funding Cycle

Because the goal of a Center is to become self-sufficient after the full ten-year funding cycle, the I/UCRC Program will not continue to fund operating Centers after ten years. This allows the limited I/UCRC Program funds to be used to establish new Centers. However, if an operating Center, in or beyond its tenth year of I/UCRC support, adds significantly new intellectual substance to its research program and continues to meet the criteria of the I/UCRC Program, it may submit a proposal for a new I/UCRC award with the same operating parameters as an initial award. Proposals to begin a new funding cycle compete against other such proposals from Centers that are beyond the ten-year funding cycle. These proposals will be subjected to a combination of individual and panel review. Awards will be based on the relative merit of the proposals and on a balance of support for both new Centers and those requesting the initiation of a new funding cycle.

Augmented Funding for Connectivity and Long Term Discovery Research

Partnerships with State Government: In order to foster partnerships between industry sectors, the academic community, and state governments, NSF may provide up to $200,000 annually per center to Centers in the I/UCRC Program that form strong partnerships with and receive financial support from state governments.

Exploratory Research Projects: Most Centers do not have enough resources to fund all projects of interest to Center members. In making the difficult decisions concerning which projects to fund, it is often the longer term or more basic research projects that are the hardest to justify for funding as the benefits of the research may not be immediately apparent to Center members. However, these exploratory projects are often the ones with the greatest potential return for Center members and the Nation in the long run. In order to foster longer term, higher risk research projects, NSF may supply up to $200,000 annually (depending upon the quality of the proposal and the availability of funds) to fund exploratory projects. These projects must be approved by the Center's industrial advisory board.

Cross-Center Collaborative Projects (Tie Projects): In order to broaden the research base of a Center, a proposal may be submitted for a collaborative project involving researchers from a Center in the I/UCRC Program with other I/UCRC or NSF funded Centers. This type of project may address industrial research interests that could not be addressed by a single Center. The experimental plan for a cross-Center collaborative project must be developed jointly by all researchers involved and must result in a single proposal submitted by all Centers collaborating on the project. Each Center may request up to $25,000 annually for two years. The funds requested by each Center must be matched by that Center, and a letter supporting the use of Center funds for this purpose from the industrial advisory board of each collaborating Center must be included in the proposal.

Industry/University Cooperative Research Fellowships (I/UF): It can be valuable for Center researchers to spend time performing research at a member company. The researcher gains a better understanding of the research needs of the company and experience with production processes in an industrial setting. In addition, the presence of the researcher at the member company is an efficient and effective mechanism for knowledge and technology transfer. NSF supports Center researchers in residence at a member company through Cooperative Research Fellowships. A proposal for an I/UF must present an experimental plan for the research to be done by the Center researcher at the member company. Funding for a fellowship is cost-shared equally with the company, with NSF funding up to $25,000.

Other I/UCRC Supplemental Programs: Centers in the I/UCRC Program are eligible for $5,000 per year to support a woman, under-represented minority or disabled undergraduate research assistant to perform Center research. Additionally, the I/UCRC Program will supply up to $25,000 per year for one or two years to support Center research performed by a faculty member from an undergraduate or predominately undergraduate institution. This proposal must be approved by the IAB, and the research may be performed either at the Center's or the faculty member's home institution.

Source: National Science Foundation, Program Announcement 97-164

APPENDIX L

U.S. TAX POLICY: RESPONSES TO QUESTIONS BY RUSSIAN OFFICIALS

Question 1: **Is there evidence that tax benefits for industrial research and development encourage industry to invest in research?**

Analyses of the effectiveness of tax benefits for industrial research in the United States have focused on the tax credit for research and experimentation. These evaluations assume that there are benefits to society from private sector research and assess the credit in terms of whether it increases research spending by companies.

Most of the early studies of the research tax credit published in the 1980s found that the credit may have encouraged additional research spending, but this increase in spending was relatively small. The research credit is effectively a reduction in the price of research. Most of the evidence from early studies found that research spending is not very responsive to price reductions. Results from those studies indicated that a one-percent reduction in the price of research would increase research spending between 0.2 and 0.5 percent. Using these estimates and information on the revenue cost of the research credit, one study estimated that during the early 1980s, the research credit stimulated between 15 and 36 cents of research spending for each dollar of tax revenue forgone.

Most recent studies indicate that the amount of research stimulated by the credit is larger than that reported by earlier research. Some studies estimate that one dollar of research spending stimulates one dollar of research spending in the short run and as much as two dollars in the long run. Some argue that these studies may provide more reliable estimates of the effectiveness of the credit than earlier studies because they analyze longer time periods and use more sophisticated methodologies. However, other studies do not support the claim that one dollar of research credit encourages at least one dollar of research spending or provide inconclusive evidence on the amount of research spending stimulated per dollar of research credit. Thus, the recent evidence is mixed.

Most of the available studies evaluate the research credit as it existed prior to its redesign in 1989. There has been little evaluation of the incentive effect of the present design of the credit, although the new credit likely increased the incentives for research per dollar of revenue cost.

A problem with providing tax incentives for research is that they are difficult for tax authorities to administer. Defining qualified research is intrinsically difficult, so that the general definition remains vague and subject to disputes between taxpayers and tax authorities. Existing evidence suggests that in 79% of the cases in which the research tax credit was audited during the first half of the 1980s, the amounts claimed were adjusted downward by nearly 20 percent. A similar incentive for research provided instead through a grant program may be easier to administer, because review and oversight would be provided by persons with specialized knowledge of a particular industry or lines of work.

Question 2: **What are the tax benefits for universities and non-profit research institutions and what is the rationale for these provisions?**

The United States tax system contains a number of provisions that give favorable tax treatment to non-profit organizations devoted to scientific research. First among these is an income tax exemption. The United States generally taxes the income of corporations. However, a non-profit corporation can be exempt from corporate income tax if most of its activities further certain purposes specified by statute. Organizations that further scientific or educational purposes qualify for an exemption. Thus, institutes dedicated to scientific research and universities that conduct research in addition to training students are generally exempt from the corporate income tax. To retain this exempt status, they may not participate in political campaigns, lobby for legislative changes to any substantial extent, or give a profit interest in their activities to private parties.

Question 3: **Is there evidence that the provision of tax benefits for universities and non-profit research institutions leads to economic pay-off?**

We are unaware of research on this specific point. However, the presence of significant benefits to society from private research is well established in the literature. The complex and variable nature of these benefits makes them difficult to measure with precision.

Question 4: **Under what circumstances are income generating activities at universities exempt from or subject to taxes?**

The income tax exemption generally does not cover income earned from conducting business activities. That income is subject to the unrelated business income tax. For example, if a university owns a factory that manufactures products for sale to the public, the income earned from the factory would be subject to the unrelated business income tax. However, the tax law gives special protection to scientific research activities. Any income that a college, university, or hospital receives from conducting research is exempt from the unrelated business income tax even if the college, university, or hospital is selling its research services or research results simply as an income-producing business rather than as a purely scientific endeavor. Similarly, if any other type of exempt organization conducts basic research—as opposed to applied research—for a profit, the income earned from the research is exempt from the unrelated business exemption. The only special requirement for this exemption is that the organization conducting the basic research must make the results of the research publicly available.

Question 5: **Under what circumstances are foreign grants to U.S. universities exempt from taxes? Under what circumstances are they subject to taxes?**

The income tax exemption generally applies to income the scientific organization earns in the course of furthering scientific purposes. If a scientific research institute receives a grant to support its research, the organization pays no income on the grant. This treatment applies to any such grant from any source, domestic or foreign. Similarly, if a large university that focuses on science receives tuition payments from students, the university pays no income tax on the tuition.

Question 6: **If universities receive gifts, such as gifts of land, are they restricted as to the use they can make of the gift?**

To provide additional support for scientific activities and other activities that benefit the community at large, the US tax code allows taxpayers to take an income tax deduction for contributions they make to nonprofit organizations that serve charitable, religious, cultural, or scientific purposes. Thus, if an individual or a business makes a contribution to a nonprofit organization that conducts scientific research, the individual or business will be able to reduce the amount of income tax he, she, or it owes to the federal government. The gift may be of cash or property. The recipient must generally devote itself to serving scientific purposes, but the tax code does not impose specific limitations on what the organization may do with the gift. However, the tax code may limit or reduce the deduction that the donor may receive depending on what the donor gives and how it is used. For example, the donor who contributes tangible personal property will get a smaller deduction if the recipient sells the property for cash than if the recipient uses the property to conduct its activities.

Source: U.S. Department of Treasury, May 1997

APPENDIX
M

UNRELATED BUSINESS INCOME POLICY OF THE UNIVERSITY OF NORTH CAROLINA

Overview

Each year the University is required to file an Exempt Organization Business Income Tax Return with the Internal Revenue Service reporting any unrelated business income generated by the activities of its academic and support units. The Internal Revenue Code states that a college or university is generally deemed to have unrelated business taxable income when it realizes gross income from any regularly conducted trade or business that is not substantially related to its educational and other exempt purposes.

Definitions

A trade or business is an activity carried on to produce income from the sale of goods or the performance of services. A specific business activity will be considered to be regularly carried on if it is conducted with a frequency and manner comparable to that of the same or similar activity by a taxable organization. Not substantially related means the activity that produces the income does not contribute importantly to the exempt purpose of the university. Any business activity conducted by a college or university primarily for the convenience of its faculty, other employees, and/or students is not taxable, regardless of the nature of the activity.

Annual Review

Each year the Controller's Office reviews all areas where unrelated business income existed or had significant potential to exist in the preceding year. All departments are asked to notify the Controller's Office of any new programs that may generate revenues that fit the definition of unrelated business income. The following are examples of potential unrelated business income generating activities.

Examples

- Any form of advertising that generates revenue for the University.
- Rental of real property if services are provided to the renter or if the property is debt financed.
- Rental of personal property (equipment, computer time)
- Rental or sale of mailing lists
- Sale of any goods or services to non-University persons or entities.

Tax Liability

The presence of these activities does not necessarily mean that a tax liability exists. It may be determined that the activity is not subject to unrelated business income tax, or if it is a taxable activity, the revenue may generally be offset by the expense incurred.

Source: University of North Carolina Accounting Services, 1995

APPENDIX N

VISITS IN RUSSIA AND THE UNITED STATES

SCHEDULE IN RUSSIA

Monday, November 10, 1997

10:00 a.m.	Opening session at Presidium of the Russian Academy of Sciences
3:00 p.m.	Group 1 visits the Institute of Synthetic Polymer Materials Group 2 visits the Institute of Organic Chemistry
7:00 p.m.	Dinner at Danilovsky Hotel with representatives of the Defense Enterprise Fund and SUN Microsystems

Tuesday, November 11

10:00 a.m.	Group 1 visits the "Lutch" Scientific and Production Association
11:00 a.m.	Group 2 visits the Mendeleyev Institute
3:00 p.m.	Group 1 visits the Institute of Steel and Alloys Group 2 visits the Karpov Physical-Chemical Institute
6:30 p.m.	Dinner at the Central House of Scientists with representatives of institutes, RAS, government agencies, and the Duma

Wednesday, November 12

10:00 a.m. Roundtable meeting at RAS Presidium

6:00 p.m. Travel to St. Petersburg

Thursday, November 13

10:00 a.m. Roundtable meeting at the St. Petersburg Scientific Center of RAS

3:00 p.m. Group 1 visits the Ioffe Physical Technical Institute
Group 2 vists the Institute of Silicate Chemistry

Friday, November 14

10:00 a.m. Group 1 visits the State Technological Institute
Group 2 visits Prometei

2:00 p.m. Group 1 visits the Joint-Stock Company Plastpolymer

3:00 pm Group 2 visits the State Technical University

SCHEDULE IN THE UNITED STATES

Monday, March 9, 1998

Seminar at the North Carolina State University (NCSU), Centennial Campus

9:00 a.m. Welcome and introductions
Mark Crowell, NCSU
Glenn Schweitzer, National Academy of Sciences
Nikolai Laverov, Russian Academy of Sciences

9:30 a.m. The Bayh-Dole Act and Its Impact on University Technology Transfer
Mark Crowell, NCSU
Francis Meyer, University of North Carolina - Chapel Hill

10:30 a.m Coffee Break

10:45 a.m. Overview of Entrepreneurial Activities in North Carolina
Bill Williams, North Carolina Department of Commerce

11:15 a.m. University Research and Global Markets Necessitate New Industrial Alliances
Robert S. Sullivan, Kenan-Flagler School of Business, University of North Carolina at Chapel Hill

11:45 a.m. Lunch

1:15 p.m. Tour of Centennial Campus
Charles Moreland, NCSU
Claude McKinney, NCSU

Tuesday, March 10, 1998

Seminar at the North Carolina Biotechnology Center

9:30 a.m. Welcome
Charles Hamner, North Carolina Biotechnology Center

10:00 a.m. The North Carolina Biotechnology Center: A State-Funded Economic Development Initiative (remarks and tour)
Doug Darr, North Carolina Biotechnology Center

10:30 a.m. Coffee Break

10:45 a.m. N.C. Small Business and Technology Development Center
Ron Ilinitch, North Carolina Small Business and Technology Development Center

11:15 a.m. Venture Capital's Role in Business Development
Dennis Dougherty, InterSouth Partners

11:45 a.m. U.S. Patent Law Provisions Which Promote University-Based Patenting and Technology Transfer
Kenneth Sibley, Myers, Bigel, Sibley & Sajovec

12:15 p.m. Lunch

1:45 p.m. Can the Science Park Concept Be Effectively Transferred to Russia?
Michael Luger, University of North Carolina - Chapel Hill

2:15 p.m. Federal Role in R&D Stimulation
Michael Stroscio, U.S. Army Research Office

2:45 p.m. Technology Development and Commercialization at Research Triangle Institute
Alvin Cruze, Research Triangle Institute

3:30 p.m. Small New Business (Case Study)
Bradley Lienhart, MiCELL Technologies, Inc.

4:00 p.m. North Carolina World Trade Center
Raymond Farrow, North Carolina World Trade Center

4:30 p.m. Closing Remarks
Nikolai Laverov, Russian Academy of Sciences
Glenn Schweitzer, National Academy of Sciences

4:45 p.m. Tour of First Flight Center
John Ciannamea, First Flight Center

Wednesday, March 11, 1998

9:00 a.m. Travel to Washington, DC

12:00 Meeting with Dr. Bruce Alberts, President of the National Academy of Sciences

12:30 p.m. Lunch

1:30 p.m. Meeting with Representatives of U.S. Organizations
Charles Larson, Industrial Research Institute, Inc.
Cheryl Cathey, NSF Industry/University Cooperative Research Centers
Marc Stanley, NIST Advanced Technology Program
Frank Thiel, An American Perspective on Developing Research Partnerships in Russia